ENERGY AND ENVIRONMENT: MULTIREGULATION IN EUROPE

Energy and Environment: Multiregulation in Europe

Edited by

PIOTR JASINSKI
Oxecon Consulting, Oxford

WOLFGANG PFAFFENBERGER
University of Oldenburg

LONDON AND NEW YORK

First published 2000 by Ashgate Publishing

Reissued 2018 by Routledge
2 Park Square, Milton Park, Abingdon, Oxon OX14 4RN
711 Third Avenue, New York, NY 10017, USA

Routledge is an imprint of the Taylor & Francis Group, an informa business

Copyright © Piotr Jasinski and Wolfgang Pfaffenberger 2000

All rights reserved. No part of this book may be reprinted or reproduced or utilised in any form or by any electronic, mechanical, or other means, now known or hereafter invented, including photocopying and recording, or in any information storage or retrieval system, without permission in writing from the publishers.

Notice:
Product or corporate names may be trademarks or registered trademarks, and are used only for identification and explanation without intent to infringe.

Publisher's Note
The publisher has gone to great lengths to ensure the quality of this reprint but points out that some imperfections in the original copies may be apparent.

Disclaimer
The publisher has made every effort to trace copyright holders and welcomes correspondence from those they have been unable to contact.

A Library of Congress record exists under LC control number: 00134480

ISBN 13: 978-1-138-74106-5 (hbk)
ISBN 13: 978-1-138-74105-8 (pbk)
ISBN 13: 978-1-315-18304-6 (ebk)

Contents

List of Contributors vi
Foreword vii

1 Reconciling EU Energy and Environment Policy
  Francis McGowan 1

2 Reforming Environmental and Energy Policies in the Economic Transition Process
  Miroslaw Sobolewski and Tomasz Zylicz 22

3 The Fossil Fuel Levy: How (not) to Save Nuclear Power
  Piotr Jasinski 50

4 The Promotion of Renewable Energy in England and Wales: the Use of the Non-Fossil Fuel Obligation
  Cathryn Ross 72

5 Energy Efficiency in Germany
  Wolfgang Pfaffenberger, Christoph Otte 99

6 Promotion of Renewable Energy in Germany
  Wolfgang Schulz 128

7 Development of Energy and Environmental Regulation in Lithuania
  Remigijus Ciegis and Vidmantas Jankauskas 150

Index 173

List of Contributors

Remigijus Ciegis and Vidmantas Jankauskas — Kaunas, Lithuania

Piotr Jasinski — Oxford, England

Francis McGowan — Sussex, England

Wolfgang Pfaffenberger, Christoph Otte — Oldenburg, Germany

Cathryn Ross — Oxford, England

Wolfgang Schulz — Bremen, Germany

Miroslaw Sobolewski and Tomasz Zylicz — Warsaw, Poland

Foreword

Over the last twenty years, the concern with market liberalization and environmental protection to some extent shifted the emphasis of energy policy away from planning techniques towards regulation. The means as well as the ends of energy policy were changing. Thus market liberalization (and associated policies of privatization and corporate restructuring) often required the evolution of an explicit system of economic regulation, backed up by the application of competition policy while environmental protection was often pursued by a variety of regulatory mechanisms (including the establishment of independent monitoring agencies). Regulation rather than ownership, taxation or direct intervention became the principle mechanism of policy. However the regulatory processes were often very different - models of economic regulation erred towards simplicity (or as much as was possible), leaving producers and consumers with considerable autonomy, while environmental models were often premised on technical standard setting, implemented in a somewhat heavy-handed way.

Moreover the contrasting regulatory styles went deeper, encompassing conflicts of principle. To put it simply, if regulation for market liberalization was concerned to reduce energy prices, the environmental perspective was concerned with increasing them (or at least with the aim of imposing the full environmental costs of energy production upon consumers to encourage them to use energy in more environmentally friendly ways). Of course many argue that such conflicts need not necessarily happen - if negative externalities are internalized into prices in an 'economic' manner then there is no clash between the goals of environmental protection and an open market. Yet in practice the goals have often clashed not least because of the very different perspectives and interests of those engaged in the policy debate on these issues.

These developments resulted in there appearing the problem of multiregulation: environmental and energy regulation (and policies for that matter) had to be harmonised, and so had to be the EU and national level. The former was partly a consequence of growing awareness of environmental problems and of liberalisation of the energy sector in which the old model of regulation through public ownership was no longer appropriate.

At the same time, the European Commission was worried that divergent national approaches to these problems would result in considerable distortions. The papers presented in this volume show how real and difficult these issues were and still are.

Coal, electricity and non-fossil energy sources are the sectors where the conflicts between different regulatory and energy policy objectives are most apparent. In each of them EU policies have attempted to address the shortcomings (whether in competition or environmental terms) and to solve any conflicts that might have appeared.

The coal sector illustrates the conflicts between different energy objectives. On the one hand, there is a significant solid fuel resource within the EU, one which arguably contributes to a degree of energy self-sufficiency. On the other hand, that resource is both relatively expensive - *vis a vis* external competitors - and highly polluting. Thus the traditional arguments for supporting the sector appear to be overruled by other concerns. Unless serious energy market disruption takes place in the next few years the steady erosion of coal's place in EU energy balances seems inevitable.

An agreement to liberalize the electricity market was eventually reached in 1996 after further watering down of the original proposal: competition was only to be introduced for the largest consumers with a gradual opening of the market over nine years. Countries were to be able to opt for either 'negotiated third party access' or a 'single buyer' system (the latter preserving to some extent the position of the single vertically integrated utilities which enjoy a near monopoly in some member states), though in both cases the different components of the market (production, transmission and distribution) had to be 'unbundled' (a separation of accounts for each component). While the agreed reforms fall short of outright deregulation, it is clear that many member states are considering (and some such as the UK and Sweden have already implemented) more radical reforms.

At the same time however, the environmental consequences of power generation have also been considered. Starting with attempts to limit emissions of sulfur dioxide in the 1980s (a measure which focused on power plants though not exclusively), the Commission has sought to encourage cleaner generation of power.

What determined how individual Member States approached various aspects of the problem of multiregulation was that their priorities differed considerably and so did the speed with which they introduced market reforms in their energy sectors. The United Kingdom definitely led the way as far as the latter is concerned, and environmental pressures seemed to have been much stronger in Germany.

The problem of multiregulation in Germany currently materialises itself in the form of two debates: on energy efficiency and on support to renewable energy sources on the one hand and on introducing energy taxation outside of the transport sector on the other hand. After a break of several years, energy efficiency is again becoming a subject matter of academic and political discussion. This time however, in contrast with the 1970s, the whole issue cannot be simply reduced to the necessity to respond to an increase in energy prices. In nominal terms, oil prices are at the same level as ten years ago, which means that in real terms they fell considerably. The renewed interest in energy efficiency cannot be explained by energy prices for consumers either, as internalisation of external costs has not yet taken place. On the other hand, the problem of exhaustible energy raw materials is not pressing at the moment because every year the amounts discovered exceed those extracted and consumed.

There is also no doubt that there exists substantial potential to develop renewable energy source in Germany, but as long as their development requires some kind of support the main question is how to make this support compatible with the increased reliance of market forces in the German energy sector. The Electricity Feed Law, which is the most effective instrument of promotion of renewable energy sources is not supported by an overall consensus nor designed to take into account a long term perspective. Thus alternative protection and promotion measures have to be considered regarding the future of renewables in the changing German energy.

In the United Kingdom, the electricity supply industry (ESI) in general and the nuclear power in particular are good examples of interactions between economic and environmental regulation. Saving the virtually bankrupt nuclear power stations could in theory lead to potentially considerable benefits for the British economy, the British taxpayers (the revenue from privatisation) and for the environment, but none of these benefits was uncontroversial, and there were substantial trade-offs involved, especially from the point of view of the effect of exploiting these benefits on the coal industry. All of this had to be achieved in parallel with attempts to introduce competition to the ESI and to establish independent regulation for that sector. As first best solutions were virtually impossible to find, the whole exercise became an exercise in expedient policy making and it is precisely from this point of view that one has to look at it.

Once the nuclear element of the fossil fuel levy has been discontinued, and the UK nuclear industry is showing operating profits there is a clear opportunity to review the functioning of renewables support in the light of this change in context. If one accepts that there is a case for government support for renewable generation on positive externality grounds then there are several financing options available to government, perhaps the most attractive of which, especially on grounds of simplicity, is that of general taxation. However, even if one sees the role of government rather as 'enabler' than 'provider' then there does appear to be a feasible market solution that could be attempted.

Just as for the Member States of the EU the issue of multiregulation has necessarily a European dimension, so the desire to join the EU adds such a dimension to the effort currently being undertaken by the countries of Central and Easter Europe which want to join. In their case, however, one must not forget that the direct context in which the problem of multiregulation is being solved is that of systemic transformation of their economies and political systems. The chapters on Poland and Lithuania well illustrate what it all means in practice.

In Poland the issue of multiregulation found its institutional form once the Energy Regulatory Authority was created, following the 1997 Energy Act. At the same time one has to remember that over the last decade Poland has been and still is a transition economy, struggling with its centrally planned history and trying to jump start market mechanism both in the economy as a whole and in its individual sectors.

Lithuania, like other former communist countries, has been facing many difficulties in restructuring from a centrally planned economy to a market related one. The move to commercialization, liberalization and introduction of private finance in the energy sector has led to a reconsideration of the appropriate industry structure and also a move to a new type of regulation. With decentralization and liberalization of the national economy the main goal of the energy policy should be: preparation of laws, setting of the priorities and control mechanisms. Energy sector regulation and supervision, management of state owned enterprises, research and development, new plants and design should be transferred to independent regulatory institutions and the private sector. The main functions of the independent regulatory institutions should be the protection of energy consumers interests and rights, implementation of consistent pricing system and the regulation of natural monopolies.

The papers presented in this volume were originally prepared for the project 'Energy and the environment, multiregulation at national and international level' (ACE Phare P96-6118-R), partly sponsored by the European Commission and presented at a conference in Warsaw, in September 1998. The editors of this book would like to thank very much all the contributors for their efforts to make their papers available to a wider public.

Piotr Jasinski, Oxford

Wolfgang Pfaffenberger, Oldenburg and Bremen

1 Reconciling EU Energy and Environment Policy

FRANCIS MCGOWAN

1. Introduction

In this chapter, we review the interaction between energy policy and environmental policy at the EU level. The aim is to consider both how conflicting objectives have emerged in EU policy in these areas and how they have been resolved. The paper does so by considering the extent to which broader objectives (such as protection of the environment and the promotion of competition) have been integrated into sectoral policies (such as energy) and identifying unresolved issues and potential sources of conflict. In effect, the traditional pattern of energy policy-making with its emphasis on supply security has been increasingly challenged by environmental and competition concerns, in the process giving a much more 'regulatory' focus to the process of policy-making. Indeed the influence of the EU is greatest where these regulatory mechanisms are to the fore. Yet these different regulatory processes and priorities may be in conflict this raises the problem of how multiple regulation operates in practice - can the formulation of an EU energy policy provide a means of reconciliation?

The paper is in three parts. The first paper considers the evolution of energy policy in the last few decades (the long term perspective is necessary to understand just how far policy priorities have oscillated), concentrating on both the integration of new concerns into policy and the growing role of regulation as a policy mechanism. The second looks more concretely at EU energy policy and its development, again taking a long term view - not only does this show how such wider concerns have been articulated in a European setting but it also draws attention to the role of the EU and how policy changes have provided opportunities for the more substantial development of EU policies in a changing policy context (not least as a result of the growing regulatory dimension, a development which privileges the involvement of EU institutions). The third section considers the principal subsectors of the energy industry, reviewing their overall position within Europe (in terms of supply, demand, environmental impact, industrial organisation) and considering how policies towards them are complementary or conflictual. Finally we conclude with some thoughts on whether and how these different objectives can be integrated at the EU level, noting recent debates on devising an EU Energy Policy.

2. Traditional energy policy concerns and the rise of new agendas

Historically energy policy has been dominated by two objectives - to maintain secure supplies of energy resources and to keep prices of energy resources as low as possible. These two objectives, apparently reinforcing but often conflictual, have shaped decisions on the portfolio of energy resources (the types of fuel and their provenance) which are used in a particular country. It is worth noting that although 'energy policy' is a relatively recent innovation (broadly since the 1950s, intensifying in the 1970s) concern over the prices and availabilities of fuels dates back much further. Moreover, governments have been prepared to intervene to meet these objectives. This intervention on occasion took the form of diplomatic and even military interventions but more commonly involved an activist approach to organising - and even owning a stake in - the energy sector. Indeed until recently, the energy sector was one of the principal vestiges of the planned economy within western societies. Energy markets were characterised by long term planning horizons, protectionism, widespread monopoly or collusion between the principal firms, with the state at the centre of many of these activities. Moreover the strategic nature of energy - and the rationale for extensive intervention - often went far beyond the supply security issue, meeting a wider range of public policy objectives (such as employment, regional development, R&D, supplier industries and macro-economic policy). While the range of objectives which governments (and producer interests) sought to meet through energy policy was arguably contradictory and overwhelming, we should not forget that during this period both prices and emissions fell. While it might be argued that these reductions were due to technical factors, defenders of the old regime might argue that those technical choices were only possible because of the institutional context and the ensuing policy preferences, priorities and processes.

It has only been in the last twenty years (and in some respects the last ten) that the preferences, priorities and processes of energy policy have been seriously redefined. Prior to then the concern with supply security - a long term concern magnified by successive oil crises and the perceived short term strength of Opec on the one hand and the perception of natural resource scarcity in the longer term on the other - had to a large extent prevailed. Increasing self-sufficiency and technical fixes were the keystones of policy, though interestingly the 1970s energy crises also prompted a close examination of the possibilities for energy efficiency and alternative (in the widest sense) sources of energy. The catalyst for change was obviously the energy shock of 1986 - when it became clear that not only were supplies plentiful but that producers were anxious to sell to a buyers' market. In that context the worries of many over supply availability were overshadowed by those seeking price reductions - the scarcity culture of the 1970s gradually eroded as did the status and credibility of those whose claims rested on that culture.

In those conditions the profile of 'new' issues rose acutely. The first major change was on market liberalisation. While there had been indications of a need to deliver cheap energy in the past these had historically been met as a result of long term technical factors (new technology, new reserves, etc.). In a sense the shift in the late 1980s was towards market liberalisation as the mechanism for low (or lower) prices. Moreover, the consumer perspective became much more important as users sought to gain the advantages of open markets over the prevailing arrangements and were no longer prepared to pay the high 'premiums' associated with supply security driven energy policies. (These shifts in interests and preferences were reflective of much broader developments of course.)

Yet at the same time, the consequences of energy production and consumption were figuring in public debates. Fears for the environment. stemmed in part from resource depletion concerns but they also related to how energy was used and which technologies were chosen. Thus issues such as acid rain and nuclear power became highly sensitive politically, reflecting a shift to concerns over the 'quality' of energy supply. This emphasis has, ironically, reinstated the claims of many of those groups who had previously based their claims on a supply security basis: nuclear, renewables and efficiency have variously been pointed out as possible solutions to the emissions problem.

Moreover the concern with market liberalisation and environmental protection to some extent shifted the emphasis of energy policy away from planning techniques towards regulation. The means as well as the ends of energy policy were changing. Thus market liberalisation (and associated policies of privatisation and corporate restructuring) often required the evolution of an explicit system of economic regulation, backed up by the application of competition policy while environmental protection was often pursued by a variety of regulatory mechanisms (including the establishment of independent monitoring agencies). Regulation rather than ownership, taxation or direct intervention became the principle mechanism of policy (Rees, 1986). However the regulatory processes were often very different - models of economic regulation erred towards simplicity (or as much as was possible), leaving producers and consumers with considerable autonomy, while environmental models were often premised on technical standard setting, implemented in a somewhat heavy-handed way.

Moreover the contrasting regulatory styles went deeper, encompassing conflicts of principle. To put it simply, if regulation for market liberalisation was concerned to reduce energy prices, the environmental perspective was concerned with increasing them (or at least with the aim of imposing the full environmental costs of energy production upon consumers to encourage them to use energy in more environmentally friendly ways). Of course many argue that such conflicts need not necessarily happen - if negative externalities are internalised into prices in an 'economic' manner then there is no clash between the goals of environmental protection and an open market. Yet in practice the goals have often clashed not least because of the very different perspectives and interests of those engaged in the policy debate on these issues. This will be seen in the next section when we look at how different objectives have been incorporated into EU energy policy.

3. EU energy policy in principle and in practice

The EC attached great importance to the energy sector: two of the three treaties on which the EC is based are specifically concerned with energy - the European Coal and Steel Community (ECSC) and Euratom Treaties concerned the coal and nuclear sectors - while the third treaty, forming the Economic Community (EEC) covered other forms of energy. In each case, the aim was to promote a common market amongst the member states while also preventing unfair competition. Even the special nature of some energy sectors - tendencies to monopoly - were covered by the Treaties. Yet there has always been a large gap between the intentions expressed in the Treaties and the outcomes in terms of European policies. The Commission's attempts to develop an energy policy of any sort, let alone one reflecting the ideals of the treaties, proved to be only of limited success as attempts to coordinate national policies first in the coal sector and then more generally came to very little (PEP, 1963; Lindberg and Scheingold, 1970).

The merger of the Communities in 1968 saw the Commission renew its efforts to develop a CEP. In its document 'First Guidelines Towards a EC Energy Policy' (CEC, 1968), the Commission noted that barriers to trade in energy persisted and stressed the necessity of a common energy market. Such a market, based on the needs of consumers and competitive pressures would help obtain security of energy supplies at the lowest cost. However, even though the Council approved the strategy, it ignored most of the Commission's subsequent attempts to enact the proposals.

The 1968 proposals marked the high point of a market-focused energy policy. From the early 1970s onwards, Commission policy proposals reflected the shift away from market making towards supply security. This change was partly a result of the development of the Community's energy balances (growing reliance on imports, mostly oil for energy needs - see Table One) and the changes in global energy markets generally. In effect the Community as a whole was vulnerable to the oil shocks of the 1970s. Yet even then member states failed to coordinate policy in an EC setting (Daintith and Hancher, 1986; van der Linde and Lefeber, 1988). As a result the Commission attempted to develop a more strategic approach to the management of energy supply and demand. The 'New Strategy' (*Bulletin of the European Communities Supplement 4/1974*), which was only agreed to after much wrangling and dilution, envisaged a number of targets to be met by 1985 (COM(74)1960) and only a few legislative actions designed to restrict the use of oil and gas. Further rounds of energy policy objectives were agreed in 1979 (to be met by 1990) and 1986 (for 1995). The 1995 objectives included a number of 'horizontal' objectives, aimed at more general energy policy concerns, such as its relationship with other EC policies. Each round sought to build on the previous one, and although in general the goals appeared to be on target, in some cases they reflected a degree of failure either across the EC or in certain member states (COM (84) 88 and COM(88)174). In each case EU Energy Policy was more concerned with the structure of energy balances than with the structure of energy markets.

By the mid 1980s, therefore, the Commission had succeeded in establishing a place in energy policy making, but it was far from being central to member states' energy policy agendas. Instead, its role consisted of information gathering, target setting and enabling activities (the latter had a substantial budget for energy R&D and promotion). While these measures ensured that the Commission had an influence on policy, they were not without problems - some of the objectives were showing few signs of achievement while aspects of the Commission's funding strategies were also open to criticism (Cruickshank and Walker, 1981). Moreover, aside from a few legislative measures, the Commission's policy had few teeth. The locus of power remained with national governments which generally chose to follow their own energy policies, resisting too strong a Commission role.

In the course of the 1980s, however, the agenda for energy policy began to change. Developments in energy markets, the attitudes of governments towards the energy industries and the overall position of the Commission in policy-making contributed to a turnaround in the concerns of EC energy policy. The new agenda rests on two broader objectives: the creation of a competition-oriented single energy market and the pursuit of environmental protection.

A key factor in the changed regime was the shift in energy markets. Prices stabilised and faltered in the early 1980s and continued to weaken until the 1986 oil price collapse. The reasons for this were more fundamental than the rows within OPEC which precipitated the fall in prices. The price increases of the early 1980s had the effect of boosting output in OPEC countries, as well as fostering exploration and production in the rest of world. Furthermore, many countries had sought to improve energy efficiency and diversify sources of energy. Economic factors - from the recession of the early 1980s to the rise of the service economy - also dampened demand. Taken together these factors led to a massive over-capacity in supply and minimal demand growth which forced down prices. The effects were not only confined to oil: gas and coal were in equally plentiful supply, while the consequences of past over-investment in electricity capacity also boosted the energy surplus. The combined effect of these developments was to weaken the scarcity culture which had prevailed among suppliers, consumers, governments and the Commission. As prices fell and markets appeared well supplied so the concerns of policy focused less on energy supply *per se* and more on the price of supply and existence of obstacles to the lowest price.

This change in market conditions made many energy policies, especially those fostering conservation or diversification from high price fuels, hard to sustain or justify. In any case, in some countries, governments were abandoning traditional approaches to energy policy. The United Kingdom was the most notable example, making an explicit move to rely on market forces for determining supply and demand (Helm et al, 1989). Shifts in policies were under review in other parts of the EC (Helm and McGowan, 1989) though these were often conceived at a less ambitious level or pursued for rather different reasons.

This changing agenda meant that the idea of an internal energy market (IEM) was once again an issue for the EC. The Commission's thinking was revealed in 'The Internal Energy Market' (COM (88) 238), a review which set out the potential benefits of an IEM and the obstacles that faced it. The IEM would cut costs to consumers (particularly to energy intensive industries), thereby making European industry as a whole more competitive; it would increase security of supply by improving integration of the energy industries; it would rationalise the structure of the energy industries and allow for greater complementarity among the different supply and demand profiles of member states. The benefits would stem from a mixture of cost reducing competition and the achievement of scale economies in a number of industries. According to the Commission, the obstacles to the IEM were to be found in the structures and practices of the energy industries. These ranged from different taxation and financial regimes to restrictive measures which protected energy industries in particular countries and conditions which prevented full coordination of supplies at the most efficient level (the latter applying to the gas and electricity industries).

In the period since the IEM document was published, the Commission has completed the programme of measures liberalising the energy industries' procurement practices, but has been unable to achieve an effective harmonisation of indirect taxation. It has also made some progress on liberalising the electricity and gas supply markets and the offshore exploration industry but very little has been achieved by way of coal industry reform. To the extent that the policy has been successful, it has been aided not only by changes in EC decision making procedures, but also by the prospect of the Commission using its powers to investigate the energy sector from a Treaty of Rome perspective.

Since 1988, the IEM has played a major role in Commission proposals on energy policy. It has, for all its problems, shifted the emphasis in EU energy policy towards a greater focus on liberalisation. However this is not the only contender for shaping energy policy. Over the same period, another element has also gained a higher profile in deliberations on the sector: The environment.

The Commission's interest in environmental issues is not new. The formal commitment of the EC to environmental policy dates from the early 1972 when, in the wake of the Stockholm conference, the Council agreed a programme of action, while some measures on environmental problems predated even this initiative. While the Commission's concerns on environment are very wide ranging covering issues such as chemical wastes, water quality, and noise pollution, the consequences of energy choices are a major part of the policy.

The importance of EC environmental policy for the energy sector has paralleled the ascent of the issue up the political agenda in an increasing number of member states. In those cases where governments have been obliged to introduce new controls on pollution, they have sought to have them generalised across the EC so as not to lose competitiveness. The best example has been the acid rain debate where the German government, forced to introduce major controls on domestic emissions from industrial and electricity plants, has pressured for similar controls in all member states (Boehmer-Christiansen and Skea, 1990). These were agreed in 1988, setting targets for emission reduction into the next century.

The emergence of the environment has given the Commission a higher profile in energy matters and another, more robust, lever on energy policy (Owens and Hope, 1989). The importance of the issue to energy policy was demonstrated in the 1995 objectives where environmental concerns were identified as a major consideration in policy. The status of environmental issues overall was confirmed in the SEA where it was given its own provisions (allowing it to enforce decisions on a majority vote). The SEM proposals also identify the need for high standards of environmental protection in the EC and this has impacted on the IEM debate.

Integrating environment and energy has not been easy for the Commission; a document on the issue was apparently the focus for considerable dispute within the Commission because of the different perspectives of the Directorates for Energy and for Environment (COM (89) 369). However, the issue which has both brought the environment to the centre of Community energy policy making and exposed the tensions between the two policies most starkly has been the greenhouse effect.

The Commission has sought to coordinate a common European response to the threat of global warming. In 1992 the it produced proposals for decreasing emissions of greenhouse gases, particularly CO_2 (COM (92) 246). These comprised four elements: programmes to encourage the development of renewable energy sources (which have zero or very low carbon dioxide emissions) and of energy efficiency, a monitoring system and a carbon-energy tax to discourage use of fossil fuels. While the Commission has pursued policies on conservation and renewables as part of a strategy for tackling global warming, the carbon tax has been abandoned. Despite many exemptions, which were included after considerable lobbying of the Commission, the proposal drew a good deal of criticism from industries and governments. Subsequent attempts to use taxation as an instrument of environmental policy in the energy sector have also been opposed (Finon and Surrey, 1996).

Certainly, the failure of the carbon tax has obliged the CEC to downplay the role of taxation in energy and environmental policy. Commission attempts to define a strategy for meeting the Kyoto obligations on cutting carbon emissions have given much greater weight to other strategies (notably on energy efficiency, renewables and transport). Meanwhile the Commission itself is pushing for a much more integrated approach to environmental policy and energy is a key target in this new strategy (Commission of the European Communities 1998).

The new agendas do not mean that the Commission has abandoned the cause of supply security entirely. In recent years it has pursued this obejctive by various means, including the management of Community activities in the event of supply disruption and the establishment of closer ties with energy producing nations and regions, but it is the ways in which it seeks to permit the maintenance of indigenous energy resources which is of most interest here (given the potential for conflict with other objectives).

A variety of other policy objectives are pursued through EU policies, including economic and social cohesion, research and development, trans European networks and *service public*. However it is around the triangle of market liberalisation, supply security and environmental protection that much of EU energy policy has evolved.

It is worth noting that, as in the national context, the shift towards environmental protection and market liberalisation has accentuated the role of regulation and as such has provided the EU institutions with more opportunities for effective intervention. While the internal market can be achieved by a variety of means (harmonisation of taxes, coordination of investments, support for networks) it is the regulatory dimension of market liberalisation which is most important, not least because of the Commission's legal powers in this area - competition rules and the internal market provisions (by contrast, the relative poverty of the EU budget makes distributive policies harder to pursue). In the environment too, there is scope for obtaining policy change through various mechanisms but the increasing powers and competence in the environmental area have again provided opportunities for intervention. Others have noted how the shift towards regulatory patterns of governance have privileged the EU institutions, particularly the Commission. Given the contrast between past policy failures and recent policy successes (relatively speaking) this seems to be borne out by the case of energy. Yet, as noted before, the goals of energy policy are often conflicting and in an EU context the question of resolving the dilemmas of multiple regulation are acute.

4. Multiple regulation and the EU energy sector

In this section we review those sectors where the conflicts between different regulatory and energy policy objectives are most apparent. We focus on three interrelated cases: coal, electricity and 'non-fossil' sources. In each one we note how EU policies have attempted to address the shortcomings (whether in competition or environmental terms) and whether and how they have resolved any conflicts.

Coal While coal was the mainstay of the European energy economy over many decades, in the last twenty years its position has been of steady and structural decline. In 1960, coal production accounted for 85% of energy production and 55% of energy supply in the countries which now make up the Community. By 1994 production accounted for only 20% of energy produced and just over 10% of energy supply. This decline reflected a restructuring of demand away from domestic and industrial markets towards power generation and away from local production towards imports: in 1960 net imports accounted for just over 5% of coal requirements; by 1994 they constituted 36% of coal supply (imports might account for an even greater amount of coal consumed were it not for the barriers to entry in a number of member states which maintain a domestic industry).

Considering the major restructuring under way in the industry and its position in the EU's treaties and institutions, the role of the Commission in coal policy has been limited. As noted, many of the attempts to develop policy from the 1950s on came to little as a result initially of a series of crises (which were largely dealt with at a national level) and subsequent concerns over the EC's vulnerability to imported sources of energy (concerns which worked to the advantage of indigenously produced coal). The relative impotence of the Commission is demonstrated by its failure to control national subsidies. The Paris Treaty all but prohibited the provision of state aids to the industry, yet such support was endemic across the Community. The Commission sought to rectify this conflict by making its approval of aid subject to conditions and ultimate phasing out (Lucas, 1977). However national support largely continued without much Commission interference for much of the first thirty years of the ECSC.

The Commission attempted to adopt a tougher policy towards the sector in the 1980s. In 1985, the Commission proposed a much more stringent set of controls of government policies including a major reduction in the level of subsidy to the industry. As the Commission noted, whereas the old rules were dominated by supply concerns, the new ones would emphasise 'the need to achieve viability.... and reduce the volume of aid even if this means substantial reductions in uneconomic capacity' (CEC, 1985a, p.130). The proposals were opposed by most coal producing countries and the Commission was obliged to accept a less ambitious policy. This policy set a framework for continuing aid until 1993 on the basis of three criteria: to improve the industry's competitiveness; to create new economically viable capacity; and to solve the social and regional problems related to developments in the coal industry. In addition, the Commission developed a more detailed procedure for approving those aids and for reviewing progress made on improving the industry's financial and economic position and for bringing into line the different forms of aid offered by member states.

If anything the Community's effect on the coal sector has been primarily felt on the demand side, where environmental policies have been to the fore. The area of Community policy which is most damaging to the coal industry's survival is, however, the environment. Attempts to combat the problems of acid rain and global warming have targeted coal as the main culprit. The principal initiative for controlling emissions causing acid rain has been the Large Combustion Plants Directive, which, by restricting national emission levels of the principal gases, effectively requires the installation of capital intensive equipment on power stations, or the use of low sulphur fuels or alternatives such as natural gas. In some countries, the effect of the directive has been to accelerate the reduction in the domestic coal industry (Boehmer-Christiansen and Skea, 1990).

However, whereas the control of acid rain can be reconciled with the use of coal (whether from the Community or elsewhere), such a bargain is much harder with regard to the control of emissions of carbon dioxide. Although all fossil fuels emit CO_2, coal emits the most. Consequently, most strategies for controlling or reducing emissions seek to restrict the use of coal: the Commission's plans for a carbon tax would penalise coal more than other fossil fuels. If the momentum behind the greenhouse effect debate continues, coal is likely to play an ever shrinking role in Community energy balances. The Commission has indicated that it is keen to encourage the clean use of coal (its research programme has subsidised various 'clean coal' technologies) but the willingness of member states to follow suit is far from clear.

The Coal sector illustrates the conflicts between different energy objectives. On the one hand, there is a significant solid fuel resource within the EU, one which arguably contributes to a degree of energy self-sufficiency. On the other hand, that resource is both relatively expensive - *vis a vis* external competitors - and highly polluting. Thus the traditional arguments for supporting the sector appear to be overruled by other concerns. Unless serious energy market disruption takes place in the next few years the steady erosion of coal's place in EU energy balances seems inevitable.

Electricity Although electricity has been the main vector for energy policy in the post war era (as the most flexible energy resource and as the consumer of all other primary energy sources), there has been little in EC policy on electricity prior to the IEM with some exceptions. There was some attempt to encourage a more diversified fuel portfolio for power generation in the 1970s and 80s: the use of gas and oil was limited in 1975 (though it was a measure that was largely honoured in the breach), while incentives for coal, nuclear and renewables were also devised.

The Commission's view of how the IEM should affect electricity has sought to balance the special characteristics of the sector with the drive to integrate and liberalise its structure. In 1989, the Commission took the first steps to creating a single electricity market. The first element of this was a revival of its pricing policy proposals aimed at increasing the transparency of electricity prices. In a review of transparency in the energy sector (CEC, 1989a), it considered the lack of publishable information on prices to large consumers as unacceptable. It sought to devise reference tariffs against which consumers across the EC can assess and compare their own prices. The measure was accompanied by moves to increase the scope for trade (or transit) and to foster investment coordination between the utilities. After nearly two years of negotiations, the Council agreed to the transparency and transit directives but not to the Commission's plans for investment coordination, committing themselves to a better use of existing agreements in this area.

The Commission's next step was to consult with governments and industry on the feasibility of greater competition. After an inconclusive series of reports, which would probably have rejected the idea of competition were it not for the support of the British government and electricity industry, a prolonged debate within the Commission took place, the result of which was a set of limited proposals for reform. These called for an extension of market access to independent power producers, distribution companies and large consumers, with the possibility of a complete market opening some time in the future. The directive was, however, drawn up as a proposal to the Council; the Commission did not use Article 90 to force the proposals through (though it had been debated). Moreover, the directive itself was framed in a way that emphasised gradual implementation, concessions to supply security and maximum national autonomy in applying the directives (COM (91) 548; Argyris, 1993).

An agreement to liberalise the electricity market was eventually reached in 1996 after further watering down of the original proposal: competition will only be introduced for the largest consumers with a gradual opening of the market over nine years. Countries will be able to opt for either 'negotiated third party access' or a 'single buyer' system (the latter preserving to some extent the position of the single vertically integrated utilities which enjoy a near monopoly in some member states), though in both cases the different components of the market (production, transmission and distribution) will have to be 'unbundled' (a separation of accounts for each component). While the agreed reforms fall short of outright deregulation, it is clear that many member states are considering (and some such as the UK and Sweden have already implemented) more radical reforms.

At the same time however, the environmental consequences of power generation have also been considered. Starting with attempts to limit emissions of sulphur dioxide in the 1980s (a measure which focused on power plants though not exclusively), the Commission has sought to encourage cleaner generation of power. Its principal failure in this regard has of course been the carbon tax but other issues have also proved controversial. Perhaps the most notable was the attempt to develop an Integrated Resource Planning approach to future power generation. The aim here was to give greater weight to environmentally friendly sources of power (renewables principally) and energy efficiency. In practice however this measure was watered down (though there are measures in the electricity liberalisation directive which encourage renewables and cogeneration).

Non-fossil sources - nuclear, renewables, conservation Growing concern over supply security and latterly the environment have fuelled an interest in non fossil fuel sources of energy in many member states. In the 1970s and early 80s nuclear power was the main focus of interest. More recently, however, the potential of renewables and energy conservation measures have been recognised.

The growth of nuclear power in the EU has been rapid but not dramatically so. Its contribution to electricity input in the EU has risen from almost zero in 1960 to just over one third in the mid-1990s. The position varies widely from country to country, reflecting the different political climates within which the industry has developed: some such as France obtain 80% of electricity from nuclear power while others like Denmark have none. The industry has been badly shaken by scandal (Transnuklear) and crisis (Chernobyl) and characterised by highly variable operating record and cost levels (Thomas, 1988). Now nuclear power is promoted less for its economic than for its environmental benefits (since it does not emit greenhouse gases). Nonetheless, not even its proponents stick to the optimistic forecasts made in the 1950s and which persisted into the early 1980s.

The EU nuclear industry is broadly composed of utilities, national authorities and fuel agencies. In almost every case the industry is predominantly publicly owned. Advanced nuclear technologies (such as the fast breeder reactor and fusion) are even more the preserve of the public sector. As part of the ESI, and given its special characteristics, the industry has not been subject to competitive pressures. Commission policy on the sector has never lived up to the expectations of the Euratom treaty. Too many countries have endeavoured to maintain autonomy over the industry. Yet the Commission has sought to sustain the industry as much for its industrial policy implications as for energy concerns (CEC, 1970b). Considerable resources have gone into promoting nuclear power and particularly joint ventures on advanced technologies: of the 1,200 million ECUs proposed for energy research in the third Framework Programme, over 75% was allocated to the nuclear sector.

The Commission's treatment of nuclear power and the IEM is separate from that of electricity, and as a result focuses less on the economics of nuclear power as a source of electricity than on the characteristics of nuclear fuel, plant and services. As in the case of coal, the Commission notes the wide disparities in policy and practice across the EC and the relative weakness of Euratom, the Treaty guiding the sector's development and the obstacles facing the development of a European and competitive market for nuclear fuels and equipment. In the first case, the long term nature of contracts for enrichment and reprocessing means that any moves towards an internal market will have to wait for their expiry (CEC 1988c).

Aside from very significant levels of research support, the Commission has scarcely addressed the nuclear issue since the mid 1980s, confining itself to reviews of the current status of the industry. The divisions between member states have persisted, effectively preventing any Community policy to emerge (through the inertia which grants almost all Community energy R&D funds to nuclear persists). There have been four developments which may allow a policy to develop in future, however. The first is the greenhouse debate. For many in the nuclear industry, the fears surrounding global warming may rekindle interest in nuclear investments though it has not so far led to a formal declaration of Community support. The second issue is industrial policy: supporters of the sector in member states and the Commission have stressed the importance of the sector as a 'high technology' sector. A related issue is the potential for rebuilding the nuclear industry in the former socialist bloc: poor safety and performance records in East Europe and the former Soviet Union have presented the European nuclear industry with new opportunities for investment and maintaining industrial capabilities (Defrennes, 1997). The final issue is, rather paradoxically, the single market. Other issue was the Commission's scrutiny of British attempts to protect the nuclear industry during the privatisation of the electricity industry. In this case, the Commission was able to limit the level and duration of support given by the British government, justifying the exception on the grounds of supply security. If the development of an internal energy market exposed more market distortions relating to nuclear energy, this mechanism could be used again.

It may be, however, that the uncertain economics and the controversial politics of nuclear power will continue to rule it out in future EU energy strategy. In that context more and more attention and resources will have to be given to the other non-fossil options: renewables and conservation.

The most established renewable energy industry is hydroelectric power which accounts for a sizeable proportion of electricity (most major sites are in use and new developments face considerable opposition). The 'new' renewables such as mini-hydro, solar, wind and wave have largely developed in the aftermath of energy crises and growing environmental concerns. While still small in terms of power contribution, they are a fixture in many utilities and their role is set to grow in most, as their reliability and competitiveness improve. The sector industries largely consist of joint ventures between governments, utilities and manufacturers.

The importance and structure of conservation industries are even harder to discern. While advisory, architectural and control systems companies (each offering ways to reduce energy consumption) proliferate, their impact is difficult to assess (given that they are aiming to help consumers forego energy usage). The overall improvement in energy efficiency must be partly due to these companies but also to other factors such as economic restructuring and price effects. Largely private, these companies have received varying degrees of support from governments while the energy industries have for the most part been lukewarm perceiving it as a threat to growth of their market. More recently, however, some large energy companies have adopted a higher profile on conservation issues as means of developing their market and diversifying.

Policy initiatives have also intensified largely as a result of the pressure of environmental concerns. On renewables, the Commission announced in 1988 a recommendation to allow favourable access for such supplies (on the basis of their environmental benefits) to public grid systems. Further measures were proposed in the wake of the Commission's greenhouse strategy. A four-part initiative - ALTENER - was tabled in 1992, involving the promotion of a market for renewables, fiscal and economic measures, training and information (Tiberi and Cardoso, 1992) while more recently the Commission published a Green Paper on Renewables and set the objective of doubling the share of renewables in the Community's energy balances (COM(96) 576). On conservation, the Commission has developed a number of programmes, the most recent of which - SAVE - sets out a variety of measures including labelling of appliances, third party finance, audits and inspections (Fee, 1992). Given parallel interests in member states, it may be that these options will be taken more seriously in the future than they were in previous energy strategies. However, an attempt to overcome the regulatory barriers to the development of these options looks unlikely to be accepted. In recent years the Commission has sought a new approach to energy investment - so-called integrated resource planning is designed to factor in 'externalities' (most importantly impact upon the environment) to investment choices. Such techniques have been used in the US and some member states to support renewable and conservation options. The measure was opposed by the larger utilities across the Community as well as by a number of governments and has since been downgraded to a 'recommendation' rather than an enforceable piece of legislation.

5. The prospects for a common energy policy

Such a variety of activities, along with the increased recourse to the Community institutions by member states and pressure groups on energy matters, would suggest that the Commission anticipated the Community finally taking responsibility for energy policy. However, attempts to formalise its role in the Maastricht Treaty were unsuccessful. While the Commission was able to insert a relatively weak commitment to a Community role which was kept in the draft Treaty up to the very last negotiations, a number of member states indicated their objections to it and obtained its removal at the last stage in the negotiations. The Commission subsequently embarked on an extensive consultation exercise in order to clarify its role in energy policy-making. A Green Paper was published at the end of 1994 with a White Paper following at the end of 1995. Both documents stressed the importance of energy matters by drawing attention to the prospect of increased energy dependence: Commission forecasts suggested that with imported energy would account for as much as 70% of energy needs by 2020 (CEC, 1995). The documents reiterated the need for a Community energy policy on the basis of reconciling the objectives of supply security, environmental protection and an internal market; a Community dimension was justified on the basis of existing treaty powers (particularly in competition policy and the internal market), the international nature of energy markets and problems and existing policy and budgetary commitments (CEC, 1995). The White Paper established an Energy Consultative Committee to ensure transparency and set out an extensive work programme for the Community in the energy sphere (though interestingly the Commission has also been willing to review, and where necessary discard, existing policies: since the White Paper the Commission has conducted two reviews of energy legislation and recommended the rescinding of some measures (including those affecting oil and gas use in such sectors as power generation).

Despite its ambitious scale, the White Paper did not seek to justify the inclusion of energy in the next round of Treaty negotiations. However, the Maastricht Treaty had included a condition that the status of energy - along with some other policies - be reviewed as part of those discussions.

In its report the Commission, while being careful not to call explicitly for a chapter on energy, indicated that inclusion was desirable given the various goals of Community energy policy and the need to rationalise the coverage of the energy sectors across the Treaties. The Commission followed this up in the latter part of 1996 and early 1997 with documents designed to justify a Community role: although these documents spelt out a range of recommendations (including a new set of energy policy objectives) they were clearly designed to strengthen the case for a formal CEP (COM (96) 431). The 1997 Treaty makes no new mention of energy, let alone a New Chapter.

6. Conclusion

It could be argued that the development of a Common Energy Policy - the Holy Grail of EU policy in this area since the 1950s - would provide the means for reconciling the different priorities of environmental and economic regulation in the EU (as well as incorporating supply security and other objectives). From this perspective, the failure to secure legal commitments in this area leaves policy in limbo. Yet is limbo such a bad place to be? If we consider the dismal record of common policies in other areas (transport and agriculture spring to mind!) then the advantages of coordination on the basis of EU law are highly questionable. By contrast, piecemeal derogations (of the sort exercised by the Competition authorities over state aids to energy efficiency and renewables programmes or to the agreement amongst car manufacturers to limit emissions) may be effective in providing a limited exemption to competition rules on the basis of a perceived higher priority. They also provide an opportunity to keep such schemes under scrutiny. While it might be argued that such *ad hoc* approaches risk departmental turf wars, it is unlikely that the latter would disappear just because of a common policy.

Appendix
The EU's energy objectives for 1985, 1990, 1995 and 2010

1985 Objectives
- To increase nuclear power capacity to 200 GW.
- To increase Community production of oil and natural gas to 180 million tonnes oil equivalent.
- To maintain production of coal in the Community at 180 million tonnes oil equivalent.
- To keep imports to no more than 40% of consumption.
- To reduce projected demand for 1985 by 15%.
- To raise electricity contribution to final energy consumption to 35%.

1990 Objectives
- Reduce to 0.7 or less the average ratio between the rate of growth in gross primary energy demand and the rate of growth of gross domestic product.
- Reduce oil consumption to a level of 40% of primary energy consumption.
- To cover 70-75% of primary energy requirements for electricity production by means of solid fuels and nuclear energy.
- To encourage the use of renewable energy sources so as to increase their contribution to the Community's energy supplies.
- To pursue energy pricing policies geared to attaining the energy objectives.

1995 Objectives
- To improve the efficiency of final energy demand by 20%.
- To maintain oil consumption at around 40% of energy consumption and to maintain net oil imports at else than 1/3 of total energy consumption.

- To maintain the share of natural gas in the energy balance on the basis of a policy aimed at ensuring stable and diversified supplies.
- To increase the share of solid fuels in energy consumption.
- To pursue efforts to promote consumption of solid fuels and improve the competitiveness of their production capacities in the Community.
- To reduce the proportion of electricity generated by hydrocarbons to less than 15%.
- To increase the share of renewables in energy balances.
- To ensure more secure conditions of supply and reduce risks of energy price fluctuations.
- To apply Community price formation principles to all sectors.
- To balance energy and environmental concerns through the use of best available technologies.
- To implement measures to improve energy balance in less developed regions of the Community.
- To develop a single energy market.
- To coordinate external relations in energy sector.

2010 Objectives

- To meet Treaty objectives notably market integration, sustainable development, environmental protection and supply security.
- To integrate energy and environmental objectives and incorporate the full cost of energy in the price.
- To strengthen security of supply through improved diversification and flexibility of domestic and imported supplies on the one hand and by ensuring flexible responses to supply emergencies on the other.
- To develop a coordinated approach to external energy relations to ensure free and open trade and secure investment framework.
- To promote renewable energy resources with the aim of achieving a significant share of primary energy production by 2010.
- To improve energy efficiency by 2010 through better coordination of both national and Community measures.

References

Alting von Geusau, F. (1975), 'In Search of a Policy' in Adelman, M. and Alting von Geusau, F. (eds) *Energy in the European Communities*, Sijthoff, Leyden.

Boehmer-Christiansen, S. and Skea, J. (1990), *Acid Politics*, Pinter, London.

Carraco, C. and Sinisalco, D. (eds) (1993), *The European Carbon Tax*, Kluwer, Dordrecht.

Commission of the European Communities (1968), 'Premieres Orientations pour une Politique Energetique Communautaire'.

Commission of the European Communities (1974), 'Community Energy Policy Objectives for 1985'.

Commission of the European Communities (1988), 'Review of Member States Energy Policies'.

Commission of the European Communities (1988), 'The Internal Energy Market'.

Commission of the European Communities (1989), 'Energy and the Environment'.

Commission of the European Communities (1992), 'A Community Strategy to Limit Carbon Dioxide Emissions'.

Commission of the European Communities (1994), 'For a European Union Energy Policy - Green Paper'.

Commission of the European Communities (1996), 'An Energy Policy for the European Energy Policy'.

Commission of the European Communities (1996), 'Energy for the Future'.

Commission of the European Communities (1997), 'An Overall View of Energy Policy and actions'.

Commission of the European Communities (1997), 'The Energy Dimension of Climate Change'.

Commission of the European Communities (1997), 'Restructuring the Community Framework for the Taxation of Energy Products'.

Commission of the European Communities (1998), 'Partnership for Integration: A Strategy for Integrating Environment into EU Policies'.

Commission of the European Communities (1998), 'Report on Harmonization Requirements'.

Daintith, T. and Hancher, l. (1986), *Energy Strategy in Europe*, de Gruyter, Berlin.

Finon, D. and Surrey, J. (1996), 'Does Energy Policy have a Future in the EU' in McGowan F. (ed) (1996).

Lucas, N. (1977), Energy and the European Communities, David Davies Memorial Institute of International Studies, London.

McGowan, F. (ed) (1990), Conflicting Objectives in European Energy Policy, *Political Quarterly*.

McGowan, F. (ed) (1996), *European Energy in a Changing Environment*, Physica, Heidelberg.

Skea, J. (1998), *Flexibility, Emissions Trading and the Kyoto Protocol*, mimeo.

2 Reforming Environmental and Energy Policies in the Economic Transition Process

MIROSLAW SOBOLEWSKI AND TOMASZ ZYLICZ

1. Introduction

The purpose of this paper is to identify and discuss problems arising at the interface of environment and energy policies in a transition economy. The process of reforming these policies in Poland is scrutinized with particular emphasis on an attempted departure from the dominance of coal towards greater reliance on hydrocarbons and renewable sources.

The authors observe that a visible progress was accomplished in environmental protection in Poland in 1989-1996 (the period covered by statistical records available). To a limited extent only this progress could be attributed to an increased economic efficiency in general and to an increased energy efficiency in particular. Steadily declining emissions were accomplished despite very modest achievements in the energy sector. This leads to a conclusion that environmental protection policies proved more effective than those aimed at energy saving in the transition period in Poland.

As far as renewable energy is concerned, the authors list specific measures that have been encouraged by the government in Poland. For the time being, these measures produced a rather limited response from the economy. This might have been caused by the fact that the new energy policy was developed at a relatively slow pace, and the new Energy Law was passed in 1997. Hence it would be premature to expect immediate results.

Having reviewed both environment and energy policies as they relate to each other, the authors discuss two initiatives that are not a part of the state policy, but nevertheless complement it. The first case study demonstrates how an independent foundation gets successfully involved in supporting projects that serve both environmental and energy efficiency objectives. The second one illustrates the potential inherent in international assistance, but at the same time shows that this potential can be used in a way that is far from effective and efficient.

2. The environment and efficiency failure of non-market economies

It has been well documented (Zylicz, 1998) that Centrally Planned Economies (CPE) used to exert a significantly greater pressure on their environmental resources than comparable market ones. For instance, there was a clear distinction between the European Union (EU) and the CPEs in Central and Eastern Europe (CEE) with respect to energy and resource intensities of their domestic products (see Table 1, below). More detailed analyses reveal that these intensities tend to decline when material welfare increases in either group. However even the wealthiest CPEs (the former German Democratic Republic, Hungary and former Czechoslovakia) had in general higher intensities than the least wealthy EU states (Portugal, Greece and Ireland). This suggests that the two groups of economies did not follow the same pattern of development.

Environmental pressure in a given country is affected by (1) the level of overall output, (2) its sectoral composition, and (3) pollution and resource intensity of the various sectors. With GDP per capita numbers much below the average for Western Europe, poorly performing former CPEs could not be considered 'victims' of their excessive output levels.

A bias towards heavy industry in national product mix and the resulting under-representation of the service sector are perhaps the most important single cause for the continuing pressure on natural resources in CPEs and their heirs.

Furthermore, there were in CPEs some additional misconceptions in environmental policy itself. The excessive pollution was often linked to an inappropriate level of taxes or fines, and thus the attention of policy-makers was distracted from the real issues. Almost never was it acknowledged that in the system where all essential inputs were allocated administratively and the firms were operating under so-called soft budget constraints there was no need to pay much attention to price stimuli. Besides, not many analysts were aware of the fact that even in the developed market economies most of the environmental improvement were achieved by means of non-economic instruments such as standards and administrative decisions.

Table 1 Economic pressure on the environment in the late 1980's

Line		CEE-6[1]	EC-12[2]
1	GDP *per capita*, $1000/person	4.8	10.1
2	- Energy intensity of GDP, - TOE[3]/$1000	0.77	0.23
3	- Energy intensity of GDP, - TOQE[4]/$1000	0.56	0.21
4	- Water intensity of GDP, m^3/$1000	153	82
	Pollution intensity of GDP:		
5	- industrial solid waste, tons/$1000	1.0	0.4
6	- wastewater, m^3/$1000	83	24
7	- gases[5], kg/$1000	51	24
8	- dust, kg/$1000	13	1

[1] Bulgaria, Czechoslovakia, GDR, Hungary, Poland, Rumania.
[2] European Community of the Twelve.
[3] ton of oil equivalent.
[4] ton of 'oil quality' equivalent: identical with TOE in the case of oil and gas; 2 TOE of coal are assumed to be 1 TOQE; 1 TOE of hydro- and nuclear electricity is assumed to be 3 TOQE.
[5] excluding carbon dioxide.
Source: Zylicz, 1998

The pure effect of policy failure - as opposed to the effects of general system inefficiency - becomes evident when comparing lines 2-7 and 8 in Table 1. In general, the European non-market economies required roughly 2-3 times more inputs to produce a given effect, and this explains much of the difference in their environmental performance. In particular, the hard-to-abate gaseous emissions resulting from fuel combustion (line 7) can be explained in terms of excessive energy use which is typically 2-3 times higher than in the West (lines 2 and 3). Energy intensities of GDP are either directly or indirectly attributable to system inefficiency. CEE countries - especially Czechoslovakia and Poland - used a fuel mix of a much inferior quality than the European average. Taking into account fuel quality differentials, the energy efficiency gap was smaller, but still existed and represented a direct effect of system inefficiency (line 3). Given their inability to compete in international markets, CPEs had to rely on domestic resources as much as physically possible. Hence their inability to upgrade the fuel mix could be viewed as an indirect effect of the system inefficiency (accounted for in line 2).

The differences in fuel quality are reflected in this analysis by aggregating various sources of primary energy in tons of 'oil quality equivalent' (TOQE). TOQE unit intends to capture this quality differential. The rates applied for various sources of energy are somewhat arbitrary, but they are not far from relations derived from market prices for a joule or kilowatt-hour obtained from such sources. Alternatively prices could have been used, but - given the heterogeneity of energy source categories - the arbitrariness would not be removed.

Based on these analyses, the excessive gaseous emissions in CPEs (line 7) can be attributed (either directly or indirectly) to general system inefficiency, and to a limited extent only could have been mitigated through better environmental protection practices. The particulate matter (dust) emissions (line 8) offer a sharp contrast to this pattern. Here the gap between CPEs and market economies was just enormous and could not be explained in terms of inefficiency. Simultaneously the abatement techniques are less costly and have been available (and occasionally applied - also in CEE) for decades. Thus it was only inability or reluctance to enforce the abatement, that allowed CEE industries to become 'dirty' in the most literal sense too. Since central-planning authorities were preoccupied with maximizing production, and particulate matter pollution did not affect output in a significant way, it was virtually neglected.

The comparison of lines 7 and 8 in Table 1 thus illustrates a major difference between environmental performance of the European CPEs and market economies. The latter did not do a perfect job with respect to SO_2 abatement and even less so with respect to NO_x throughout the 1980s. (Even though it would have been impossible for Poland and other CPEs to meet the requirements of the First Sulphur Protocol under the old political systems (Zylicz, 1988). Consequently the contrast between both groups of countries was not so apparent as in the case of particulate matter emissions. By no means could these be explained in terms of (excessively) consumed energy. Exactly because of firm environmental policies the developed market economies, in the 1980s, could start to enjoy low levels of dust emissions which indeed are fairly easy to abate. Despite moderate abatement costs and moderate level of technological sophistication, CEE economies failed to solve their dust emission problem clearly because of the lack of adequate policy measures.

3. Environmental performance in the economic transition process

After 1989 Poland improved its environmental performance considerably. Table 2 (below) illustrates some key aspects of the environmental policy effectiveness of the Polish economy. Table 2 follows a similar row structure as Table 1, but numbers may not be comparable due to a wide margin of error inherent in GDP estimates for non-market economies from the 1980s. In Table 1 Purchasing Power Parities were used to approximate the level of economic activity, while the subsequent table apply nominal currency exchange rates. Moreover, Table 1 uses dollars from different periods (1987-89) while the other one does from 1992. Therefore dynamics observed after the 1989 will be the focus of the analysis.

The first observation is that improved environmental performance cannot be fully attributed to declining GDP. Air pollution intensity decreased both in 1989-92, when economy was shrinking, and in 1992-96, when economy was expanding. Water use, water pollution, and solid waste intensity were more or less stable in the first period, but they improved substantially in the second. Energy intensity improved as well, although partially due to the higher quality fuel mix achieved after 1992, and to some extent due to other causes (such as changing GDP mix and technologies). The standard energy intensity of GDP achieved in 1996 is 43 GJ/USD or roughly a ton of oil equivalent per 1000 USD. This is a quadruple of a Western European level from the late 1980s, but taking into account that the purchasing power of the Polish currency is about twice its nominal value, the energy efficiency gap is 'only' 2:1. Moreover, if one takes into account the specific fuel mix dominated in Poland by solid fuels, the energy intensity achieved in 1996 is 28.54 calculated GJ/USD or 0.682 tons of 'oil quality equivalent' per 1000 USD. After adjusting for the purchasing power of the Polish currency, it indicates the energy efficiency gap vis à vis Western Europe of 'only' 1.5:1. In other words, further improvements will be increasingly constrained by the fuel mix.

The second observation becomes apparent when comparing Table 1 and Table 2. Even though the falling emission of gases (mainly sulphur dioxide) are visible, the most impressive improvement relates to particulate matter pollution. As indicated earlier, this was where the CPEs offered a striking contrast with Western European economies, and where the environmental policy failure of the former was best revealed. At the same time this is where the most spectacular improvement can be demonstrated thus suggesting that the effect results from policy reform rather than merely from economic processes.

Table 2 Environmental resource use and pollution in Poland

Year	Water withdrawal [10^9 m^3]	Waste water[1] [10^6 m^3]	Gaseous[2] emissions[3] [10^9 kg]	dust[4] emissions[3] [10^9 kg]	Industrial solid waste [10^9 kg]
1989	15.10	4406	5113	1513	170.9
1990	14.25	4115	4115	1163	143.9
1991	13.27	3754	3552	923	128.3
1992	12.57	3461	3155	684	121.9
1993	12.27	3151	3001	599	120.5
1994	11.98	3183	2941	529	120.9
1995	12.07	3020	2785	432	122.7
1996	12.01	2914	2672	391	124.5

[1] requiring treatment (i.e. without cooling waters).
[2] excluding carbon tax.
[3] from registered stationary sources.
[4] total particular matter.

Additionally, the tables offer a perspective at the relation between the allocative efficiency and environmental performance. Substantial as they are, the lowered energy and water intensities of GDP (by 11%-16% and 24% respectively), are nevertheless less striking than lowered water- and air-pollution intensities (37% and 50%-75%, respectively). This reinforces the argument developed earlier and demonstrates that the environmental protection failure in pre-1989 Poland was not just a matter of a resource intensive economy, and that post-1989 policies triggered improvements much higher than implied by increased allocative efficiency (exemplified by energy and water use). The table also suggests that the least successful environmental policy domain is industrial waste disposal where the pace of reduction (30%) lags behind the air and water improvements. The reason why solid waste management was less successful than other areas of environmental protection is not clear. Relative under-staffing of environmental authorities (visible in the Ministry of Environment at least) combined with relatively lower alertness of broad public opinion might have contributed to this outcome.

Between 1992 and 1996 the composition of GDP changed visibly. The contribution of trade increased from 13% to 14.7% while that of agriculture (with forestry) decreased slightly from 6.6% to 6%. The contributions of industry and construction decreased more visibly - from 34% and 7.8% to 27.1% and 5.3%, respectively. Aggregate industrial output grew by 41.7% with extractive industries stagnating (2.2% growth), and manufacturing expanded by 53.8%. Structural adjustments were thus an important factor in decreasing emissions, but they could not explain the actual scale of the process.

Overall pollution intensity of Poland's economy has fallen sharply not because of the declining energy use but rather despite of the moderate improvement in the average energy intensity. This should be emphasized since it contradicts early recommendations made by many analysts who expected environmental improvements to be led by advances in energy use (e.g. The Economist, 1992). This opinion became a sort of a popular wisdom, and was repeated many times and rarely opposed (Kaderjak, 1997). Very few authors acknowledge that even a drastic (inconceivable politically) coal tax of 100% would yield minor aggregate energy savings and consequently a minor emissions decline (e.g. Toman et al., 1994). Emissions of some pollutants are indeed likely to decrease following higher energy efficiency and switching to higher quality fuels. This however will be a long process and by no means should it preempt taking policy steps to contain pollution from existing processes, where cost-effective.

As in Table 1, the energy data are fairly certain and they reflect fuel use statistics which has been well established. Pollution intensities are based on emissions from registered stationary sources whose share in total emissions is different for various pollutants. These shares, however - ranging, for instance, from an estimated 40%-60% for dust, to 65%-70% for sulphur dioxide - have been fairly stable over the period analyzed. Hence taking into account such uncertainties may weaken previous conclusions somewhat but it cannot invalidate them.

4. Environment *versus* energy policy after 1989

In their struggle for a friendlier, safer environment, economies in transition have to overcome two barriers: general system inefficiency and inadequate policy principles. Economic reforms gradually remove the first obstacle but preventing social disturbances or dealing with their effects may postpone achieving efficiency goals and make environmental improvements slower. Consequently environmental policy makers cannot take the beneficial results of market reforms for granted.

The other barrier is not easy to overcome either. Quite commonly in CPEs, the failure of environmental policies was linked to an inadequate level of charges and to the lack of developed markets. Thus there was a natural temptation to conclude that with the advent of a market economy some 'optimally calculated charges' will - at last! - start to work. Waiting for positive spillover effects of emerging markets has been the main cause of the lack of environmental improvement in some economies in transition. Poland may well have ended up in this group if the government yielded to numerous advocates of 'liberal' environmental policies. Recommendations to build these policies on pollution taxes were coming both from politicians and academics. In the latter case the advice was based on poor scholarship and ignorance of the Western experience. In the former case motivation must have been shrewder. The argument to let market forces determine the level of abatement, once externalities are accounted for by the taxes, could be appealing to liberally-oriented administrators who ran the government in 1989-1991. The Machiavellian point was, however, that the Minister of Industry (a promoter of the idea) perfectly knew that throughout the 1990s, any attempt to unilaterally elevate pollution taxes to their biting levels would fail for political reasons. Hence the recommendation 'not to mess with market forces' was in fact equivalent to admitting that environmental protection was premature and should be postponed until later.

That the environmental *laissez faire* solution was a real option in Poland in 1989 one can see from what happened to a number of other sectoral policies. As a result of a liberal ideology, tariff duties were unilaterally lowered which left domestic producers unprotected until the government had to yield to their complaints. No housing policy was adopted in anticipation that the market would solve the housing shortage inherited from the past. Likewise the development of energy, transport, industrial, and agricultural policies was postponed until much later. Given that all sectors of society and economy were in deep disequilibrium at the beginning of the transition period, most of such 'deregulation' policies were doomed to failure.

The environment was addressed by the Polish government in a pragmatic way. In 1990 the Ministry of Environment drafted its major policy document (Ministry, 1990a), which was quickly approved by the Council of Ministers, and then officially accepted by the Parliament in 1991. Its section on policy implementation was based on a more detailed analysis of economic instruments (Ministry, 1990b) serving as a technical appendix to the main document. This Outline of economic instruments contained several specific recommendations and it was referred to in a number of subsequent political debates on various pieces of environmental legislation. (It was circulated in Polish and English versions to facilitate both internal and international consultations.) Some of the recommendations have been implemented since that time. Some - most importantly those to be included in the new Environmental Protection Act (whose first draft was prepared already in 1991) - are still waiting for final consideration. Yet others are being occasionally challenged by alternative proposals submitted to the Ministry or originating from the Ministry officials (Zylicz, 1998).

According to the Outline, economic instruments perform a number of useful functions. Most importantly they help to minimize overall costs of environmental protection through an efficient and non-arbitrary differentiation of control requirements. The document envisaged a substantial role both for marketable permits and pollution charges, the latter, however, serving mainly revenue raising purposes. While the document proved successful at arguing for substantial pollution charge increases (Zylicz, 1994a) (the charge revenues now correspond to .5% of GDP and provide more than a reasonable public share in environmental project financing) it failed to speed up the introduction of marketable permits.

Poland's environmental policy statements of 1990 were the first among those adopted by governments in economies in transition. By 1993 all countries had some environmental programmes. As a rule, they were not very specific with respect to the choice of instruments. None of these programs was also selective enough to provide for clear enforcement and investment priorities. The first policy adopted in Poland was fairly selective, but it was subsequently amended several times in order to accommodate various political concerns. The World Bank (1993) tried to inspire governments to rationally adopt priorities and sharpen investment criteria, but its approach has not permeated actual policy making processes.

Apart from pollution charges, emission standards seem to be the focal point of many discussions in the Ministry of Environment. The latter are sometimes perceived as the 'modern', 'Western' approach to solving environmental problems. There is thus a serious risk that Poland - together with other CEE countries - will make an unfortunate attempt to replicate the strictest (e.g. German) emission standards. The effects would be twofold. First, as such standards will be doomed to violation by many polluters for a number of years, they will contribute to the attitude of disregard for the law rather than providing incentives to abate or innovate. Second, by inflating the demand for sophisticated abatement technology which is unavailable domestically, they will affect the import intensity of the domestic production.

Most CEE countries adopted emission standards similar to those existing in the EU. In Poland fairly stringent air emission standards for new sources were established in 1990. Corresponding standards for existing sources were scheduled to come into effect in 1998. Western-like water effluent standards were introduced in 1995. In all cases there have been strong pressures exerted by polluters to relax these requirements. The Polish Ministry of Environment sees two ways to accommodate justified economic concerns without abandoning the standards. One is emission trading while the other is compliance schedules, i.e. 'contracts' between polluters and inspectors allowing the former to exceed permissible emissions if steps are taken to comply in a specified time frame. Both instruments are provided in draft Environmental Protection Acts and both were practiced 'on an experimental basis' after 1989.

Despite legal shortcomings and insufficiently selective priorities, Poland's environmental performance has been fairly good since 1990. It should be stressed that, by and large, improvements were achieved as a result of conscious policy measures rather than economic adjustments. In particular, the emissions of many pollutants declined much faster than implied by improvements in energy efficiency and fuel mix.

One can say that little changes affected the Polish energy sector. As before, the fuel mix is heavily dominated by coal (78% in 1989 and 75% in 1996). 'Non-fossil' energy - including renewable sources - accounted for 1.6% in 1989 and for 4.8% in 1996. This three-fold increase was to a large extent caused by changes in statistical reporting when, in 1993, the registered amount of peet and wood quadrupled. Still, however, the contribution of these so-called non-conventional sources has remained marginal.

The government of Poland developed its energy policy much more slowly than the environmental one. Draft energy policy statements were rejected by the Parliament as inconsistent with accepted environmental policy, among other things. Nevertheless a new Energy Law was finally passed in 1997 (thus before the new environmental laws which are still pending). This important regulation is expected vitalize the country's energy policy, promote efficiency, and serve sustainable development purposes.

The Law requires that the government develops a 15-year outline of energy policy and reports to the Parliament on its implementation periodically. The outline must take into account sustainable development principles (article 15). In particular, it is supposed to indicate what environmental protection measures are anticipated, how non-conventional (including renewable) sources are promoted and how energy savings (especially thermal insulation) are encouraged.

The energy sector is regulated by the government through the Energy Regulation Authority. The Authority grants concessions required to operate in the sector, and controls prices of gas, electricity, heat and lignites (sold to power plants). Operations in hard coal extraction and distribution are not subject to these regulations. The concessions extend over the period of not less than 10 years and not more than 50 years. Except for solid fuel suppliers, all major energy producers or distributors need a concession to operate. Exempted from this regulation are sources below 1 MW and distributors below the flow capacity of 1 MJ/s. As well exempted are low-voltage (below 1 kV) electricity grids, gas retailers with annual sales below 25,000 ECU, and liquid fuel retailers (article 32). The Authority approves prices taking into account operators' costs which, in particular, may include: environmental protection, energy saving investments implemented at customers' facilities, and developing non-conventional sources. These provisions act as *de facto* tax allowances.

Non-conventional energy is favored by the Energy Law through a provision requiring distributors to buy the electricity or heat generated from such sources (article 9.4). Further details of this regulation are specified in an ordinance of the Minister of Economy.

The establishment of the Energy Law has been considered by policy analysts as a compromise between a demand for making the energy policy consistent with the environmental one and preserving a strong political position of domestic producers and distributors. The final outcome has not been determined yet, as many specific regulations stipulated by this Law have not been issued by the late 1998. For instance, it is too early to judge how the provisions favoring the energy saving and non-conventional fuel projects will affect the domestic demand for coal. From an economic point of view, this demand is a factor making the performance of the energy sector more environmentally expensive. On the other hand, from a social as well as strategic security point of view, the abundant deposits of hard coal are seen as assets rather than liabilities.

There is no inherent conflict between keeping an economically viable hard coal production and eliminating the most environmentally troublesome part of the coal combustion. It is estimated that 37% of the total particulate matter pollution originates from households and small district heating stations. These are low-stack emissions leading to immediate and acute human exposure. For technical reasons these sources cannot apply effective abatement installations. At the same time, they use the very worst quality of coal prompted by its easy availability at low prices. The problem is caused by approximately 5 million tons of coal, i.e. less than 5% of the total domestic coal use of 110 million tons in 1996. Eliminating the lowest (in terms of quality) 5% of the coal supply would largely solve the environmental problem without wiping out the coal industry.

Several ways to free the Polish settlements from the plague of low-stack emissions were contemplated. Pollution charges were once considered an option, but the current market structure precludes such an option. As combustion sources are dispersed, it would be impractical to levy and enforce emission taxes. Hence the taxes should have been levied in the form of product charges added to the wholesale price of coal. The mines, however, operate under cartel-like agreements which apply cross-subsidies and regulate prices so that the worst quality varieties of coal are sold much below extraction costs. As a result, any hypothetical 'corrective' product charge should at least double the price in order to make the dirty coal as expensive as the clean one. Even though theoretically possible, this is an option seen by most specialists as politically unfeasible.

A more feasible alternative is to promote fuel switching and similar measures by subsidies rather than negative incentives. The new Energy Law encourages energy suppliers to cross-subsidize such measures from their operating budgets but it is too early to assess its effectiveness in this field.

5. Poland's policy towards renewable energy sources

It is generally agreed that the development of power industry based on renewable sources may be a good solution for a number of environmental problems caused by the power sector in Poland. The promotion of the sector in question, being in line with the principles of sustainable development, is one of the priorities of the State Environmental Policy (PEP, 1991). There are numerous benefits from the renewable power industry development. These include:

- lower emissions of such pollutants as CO_2, NO_x, SO_2 (which is crucial for improvement of the state of the environment as well as for the ful-

fillment of Poland's international obligations resulting from the relevant ecological conventions);
- Increased fuel diversity and energy security of the state;
- rationalization of economy with natural resources and potentially positive influence on the State's trade balance (being an outcome of limited demand for imported energy sources);
- job creation, enhancing local labour and service markets;
- Despite this positive characteristics, the renewables' share in the fuel mix used by Polish power industry is only slightly over 1%. In this respect, it seems important to outline the prospects of utilization of the renewable energy sources in Poland as well as identify the main obstacles in the development of this sector.

6. The state of renewable energy sector in Poland

Assessments of the potential (physical and technical availability) of renewable energy sources existing in Poland are not consistent. Some studies point out that the resources in question exceed the total demand for energy of the country (even if one takes into account solely the resources that can be exploited with the use of current technologies). Other publications, however, consider the renewable energy sources' potential as marginal. The latter include the government's document which defines the assumptions of Poland's power policy till 2010 (Guidelines, 1995).

A report published by the Polish Ecological Club (PKE) claims that in Poland the potential of such resources as direct solar energy, biomass energy, geothermal energy and - to a smaller extent - water and wind energy should not be neglected. According to that study, the country's technological potential amounts to 3860 PJ and the exploitation of it at the rate of 620 PJ/year would allow to cover around 15% of the commercial energy demands. This will not be achieved, however, without a significant financial involvement of the State in subsidising and promotion of the renewable power industry. A much more moderate account of the possibilities of the renewable energy sources utilization is given in another study (Strategies, 1996); this estimates technological potential of renewable energy in Poland at 337,2 PJ and forecasts its actual use in 2030 at 161 PJ/year, with individual sources contribution such as: large hydropower plants - 15,8 PJ/year, small and medium scale hydro power - 2 PJ/year, wind power plants - 1,9 PJ/year, geothermal - 40 PJ/year, air and water solar installations - 41,2 PJ/year, biomass (thermal installations using wood waste and straw, biofuels, biogas) - 60,1 PJ/year. Official documents concerning renewable energy (MoIT, 1996) estimate its potential share at 6,6% of TPES by 2010 (in the most optimistic development scenario, including large hydro projects).

For the time being, the main sources of renewable energy used in Poland are biomass combustion (approximately 55 PJ/year), hydropower plants which operate within the country's energy system (approximately 6 PJ/year) as well as small water power plants. Obtaining energy from other renewable sources (wind, solar, geothermal energy and biogas) is still of minimal importance. While none of the different renewable energy sources has a particularly large potential in the whole country, the regional potential of some carriers can be quite high.

A short characteristics of the current state of the renewable energy sources exploitation in Poland is given below.

6.1. Hydro Power

The energetic potential of the Polish rivers is estimated as 13,2 TWh/year; this includes 1,1 TWh/year for small power engineering. There are 18 big hydro-power stations operating in Poland (over 5 MW each), which provide electric power in the amount of approximately 4000 GWh/year and 200 small power stations with the total capacity of approximately 85 GWh/year. The generating capacity of the Polish hydro-power stations exceeds 2000 MW (including 1366 MW of pumped-storage power stations, which are used to meet peak demand and are not unanimously recognized as renewable sources). It is estimated that it is possible to build about 700 small hydro-power plants making use of the already existing dams and additional 400 stations next to projected artificial dams and water reservoirs built for agricultural purposes. Some analyses suggest that currently only 15% of the economic potential for hydropower in Poland is used.

6.2. Geothermal

At present, two geothermal power plants operate in the Podhale region and in Pyrzyce (near Szczecin). The third enterprise of this kind is currently under construction. In Poland, thermal waters are exploited only as a heat source; they are not used for electricity generation. Their resources available for practical utilization are estimated at 117-263 PJ per year. They are located mostly in the lowland region, especially in the area between Szczecin and Lodz and in the regions of Grudziadz and Warsaw. However, taking into account their high mineralization, low efficiency and deep location, not all resources can be efficiently used for power engineering purposes.

6.3. Solar Energy

The geographic location of Poland gives potential conditions for solar energy exploitation, especially in agriculture. The density of solar radiation energy approximates 980-1100 kWh/m^2 per year. Average annual insolation is 1600 hours. In some regions of the country passive solar installations are used. Air collectors are most often used in agriculture for drying crops. They are used for an average of 300-600 hours per year. Liquid collectors are mainly used for heating water in apartments, camping and summer houses, sports and recreational buildings, stock buildings, feed stores as well as for heating water in reservoirs, basins and technological water in small industrial plants. Photovoltaic cells which convert solar radiation to electricity are hardly ever used in Poland.

6.4. Wind energy

For the time being, there are ten wind power plants operating in Poland (their nominal power is 95-1200 kW), which supply electricity to the grid. For example such power plants work in Lisewo and Swarzewo (Gdansk region) and Rytro near Stary Sacz (Karpaty Mts). From the technological and economic point of view (the annual average speed of wind over 4 m/s) it would be reasonable to exploit wind power plants at the 1/3 area of Poland. The prime locations for installation of wind turbines are at the Baltic Sea coast, in the Suwalki region as well as in the South-Eastern region of Poland. Apart from commercial wind power stations, there are some one hundred small autonomus wind power plants not connected to the grid. The total capacity of all wind turbines installed in Poland amounts to approximately 3 MW.

6.5. Biomass

Biomass combustion has the greatest potential among all renewable energy resources used in Poland. Wood and wood-waste combustion for heating purposes in private households is common in rural areas throughout Poland. Fuel wood resources - forests, orchards and wood industry wastes - occur in all regions of the country; it is also estimated that around 1/3 of domestic straw production could be used for heating purposes. Almost all organic waste products in agricultural production, especially those coming from the animal production, such as manure, can be used for biogas production, based on methane fermentation process. There are several installations (often experimental) for biogas production in Poland. They produce heat for local needs. Biofuels: ethanol and rapeseed oil which are added to fuels, can also be treated as a renewable energy source. Biofuels might represent an attractive energy source for Poland which imports almost all of its liquid fuels. Additional environmental benefit of biofuels lies in a fact that their production allows for recultivation of contaminated soils, which should not be used for food production. Currently, ethanol produced from grain and potatoes is being used as an additive (up to 5%) to regular gasoline in some Polish oil refineries. Rape methyl ester, 'the bio-diesel' is produced in a pilot plant in Mocholek. Larger scale production of this fuel is economically unattractive. In EU countries biofuel production is supported as a means of decreasing the subsidized overproduction of other crops. The future extension of EU policies into Poland might create additional incentives for development of biofuels sector in Poland.

7. Conditions concerning renewable energy utilization

In many countries renewable energy is supported by the state. To this end, special regulations for promotion of its development are passed. In OECD countries the most frequently used methods include:

- economic and fiscal incentives through subsidies, grants or tax breaks (or, conversely, disincentives such as carbon tax on competing fuels or full cost pricing);
- the regulations which guarantee the market for renewable energy (often at favourable prices comparing to the conventional energy);

- subsidies and state support for the research and development on renewable energy technologies;
- information and education campaigns to increase awareness of renewable energy technology performance and availability;
- regulations and standards on energy use, environmental performance and energy installation siting;
- national targets, i.e. quantified plans for renewable energy development either by source or by sector;
- voluntary actions, generally between governments and industries or utilities;
- green pricing, whereby customers choose to pay more for 'green' energy.

In Poland, despite several declarations of the State's support for renewable energy, included in the official government policy documents (e.g. Energy Policy Guidelines for Poland until 2010, State's Environmental Policy), the list of specific decisions in support of renewable energy sources exploitation is not very extensive. This is a result of the government's policy assumption, according to which "generation of power from renewable sources cannot be subject to different laws than generation of power based on conventional sources"(MoIT, 1996). A majority of the policy issues surrounding renewable energy development may be viewed in three, to a large extent overlapping, dimensions: legal, economic and institutional. They are dealt with in the following sections.

8. Legal conditions

General legal conditions concerning the exploitation of the renewable sources of energy in Poland result from the laws regulating commercial activity. The issue in question is regulated in greater detail by the new Energy Law passed in 1997 (*Journal of Laws*, No 54, item 348). The latter states that the Minister of Economy is responsible for the development of the power industry sector on behalf of the government, while the chairman of the Energy Regulation Authority (ERA) is responsible for the market regulations (licensing, prices). The main aim of this law is to create conditions for guaranteeing state energy security understood as meeting current and future energy demands of consumers with ensurance of environmental protection, economical and rational energy consumption, competition development as well as protection of consumers' interests and minimizing the costs. According to this act, power enterprise performance, which covers energy generation, conversion, distribution and transfering as well as trade, requires receiving the ERA licence (in the case of heat production such licenses are to be granted by regional authorities). This requirement is not applicable to sources smaller than 1MW. Licensee is obliged to set energy prices with approval of the ERA; this aims to ensure non-monopolistic behaviour of the enterprises.

The problem of environmental protection, especially as regards renewable energy sources (which are defined in the above act as 'the sources which use in the process of conversion unaccumulated solar energy in various forms including, most of all, energy of rivers, wind, biomass energy and solar radiation energy converted in solar batteries') is treated marginally in the Energy Law. In spite of this, there are certain provisions of the act which may contribute to the increase of renewable energy utilization.

According to article1, point 2, the necessity to take into consideration the requirements of the protection of the environment in power engineering activity is one of basic aims of the act. It is also claimed that *'The Energy Policy Guidelines'* endorsed by the Council of Ministers in the form of a decree, have to relate to activities in the field of environmental protection and use of non-conventional energy sources (article 15). Article 39, in turn, defines the scope of actions required to be granted the licence for energy activity. It also claims that a licence may (although does not have to) define obligations concerned with environmental protection during or after the termination of the licensed activity. The article 40 states that granting a licence may depend on securing claims by a licensee (which may occur as a result of ecological damage caused by actions of a licensee). The licensing authority has the right to withdraw the licence if actions carried out by the licensee violate the conditions stated in the licence. The latter rule can be applicable also to environmental offenses.

The regulations concerning energy prices are also of crucial importance for environmental protection issues. The law stipulates that market prices will be introduced not later than 24 months after the law comes into force. It is assumed that the process of introducing market prices for electricity will start on 1 January 1999. One may suppose this will bring positive environmental effects, especially in light of article 47 of the law, which states that rates for gas fuels, electric and heat power should allow for covering the costs of enterprises' performance also as regards environmental protection. Moreover, it is agreed that energy rates can include the costs of undertakings which aim at reduction of consumers' energy consumption (as an alternative for building new energy sources), co-financed by power engineering enterprises, as well as actions related to alternative energy development. Additionally, the new Energy Law explicitly mentions the right of energy enterprises to differentiate tariffs according to varying costs of deliveries (article 45). The regional variation of energy prices to final consumers may increase penetration of renewable technologies in remote and rural areas where population density is low.

Interestingly, the legal solutions that favour renewable energy sources are introduced primarily in order to facilitate the process of developing local energy markets. These include, *inter alia*, an instruction to take local energy resources into account while planning the local heat supply, as well as the regulation waiving the licence obligation for fuel production and generation of power not exceeding 1 MW (which benefits small producers of the renewable energy sector).

The rule of Third Party Access (TPA) introduced by the Energy Law allows to break the monopoly of conventional energy sector and eliminates a serious obstacle for independent producers. Access to the grid, which is limited only by the need to meet appropriate technological requirements, is a crucial factor for profitability of investment in the power sector. This is also applicable to the producers who use water or wind power. Access to the grid does not automatically mean, however, that power enterprises will have to buy the energy supplied by independent producers. In this respect, the Energy Law states only that the Minister of Economy may (but does not have to) impose an obligation on power companies to purchase energy from renewable sources. Despite the fact that the act was passed 18 months ago, such an obligation has not been introduced yet. The draft of such a regulation is currently being prepared by the Ministry of Economy. However, the parties involved claim that 'the process of reaching agreement is hindered because of some hesitation on the part of conventional power sector lobby' (Romaniszyn, 1998). This regulation would be an essential condition assuring that the distribution companies could be forced to allow for some share of renewables among the electricity they sell, even if they have to pay a higher price for it.

The most important regulations resulting from the Energy Law, which can stimulate renewable energy sector development in a long term horizon are, among others:

- regulations which lead to rationalization of energy prices, in consequence of which the external costs (environmental pollution costs among others) may become included in the price of energy. For the time being, they are not fully reflected in energy prices, which gives a competitive advantage to the traditional power sector. Altering this situation can contribute to the increase of economic viability of renewables;
- regulations which oblige the Council of Ministers to consider the issue of alternative energy sources development while working out the assumptions of state power policy;
- regulations which permit the Minister of Economy to oblige the energy distribution companies to buy electric energy and heat from the renewable sources;
- regulations which allow to take into account the expenses for renewables sector development in energy prices rates presented for approval by the ERA.

The regulations which exceed the limits of the Energy Law are also to be decisive for the renewable energy development. These include the regulations and standards concerning environmental protection. The need to fulfill both the priorities of the State's environmental policy, and relevant international agreements may turn out to be one of the key factors working in favour of renewable energy. The necessity to limit the emissions of such pollutants as NO_x, SO_2, CO_2 would mean that standards specifying acceptable emission levels will gradually become stricter, which will have an impact on the costs of generation of fossil fuel based energy. As the costs of traditional energy generation increase, it will be easier for alternative energy producers to stand up for a market competition. Paradoxically, however, it turns out that strict ecological standards can also be an obstacle for the technologies which exploit non-fossil energy carriers. This can be proved by the example of norms pertaining to the allowable level of nitrogen oxides emission to the atmosphere. In the case of the latter norms, the acceptable emission rates vary depending on a kind of fuel and type of equipment using a particular fuel. As far as biomass combustion is concerned, the permitted emission is 50 g/GJ of energy the fuel contains. This norm is a few times more severe than in the case of coal-fired installations (95 - 170 g/GJ) and, according to specialists, taking into account the current state of technological development, it hampers utilization of biomass for energetic purposes in Poland (Wisniewski, 1998).

9. Economic incentives for renewable energy development

At the beginning of the 90s individuals engaged in commercial activities and farmers using renewable sources of energy in rural production were exempted or granted concessions from income and farming tax for utilization of renewable sources of energy (exemptions from income tax for 5 years for selling energy in case of the former and rebates in farming tax up to 15 years for capital investment in case of the latter). In 1993 Minister of Industry and Trade issued a decision guaranteeing premium rates for electricity produced from renewable sources (up to 5 MW), assuring a selling price for small hydro and wind operators at 85% of the final consumer price. This provision is still applicable, although there are plans to revise it in a close future.

Apart from the regulations mentioned above, for the time being there are no system mechanisms concerning financial support to investment projects in the field of renewable energy sources development. The State's activity is limited solely to granting loans from environmental funds and the EcoFund. In the years 1990-96 over 720 projects were financed from these sources, at the total sum of approximately 10 mln USD (Nat. Com., 1998). According to the analysis conducted as a part of a research programme devoted to working out the strategy for reducing the emission of greenhouse gases in Poland (Strategies,1996), the minimum financial support which will ensure profitability of investments in the renewable energy sources sector has been assessed at:

- between 50% to 90% of capital investment for autonomous wind power plants, photovoltaic cells and biogas works;
- approximately 30% in the case of passive solar installations (liquid and air collectors), agrorefineries and geothermal power plants.

Small water power plants will be profitable at the loan interest rate lower than 10%. Only technologies of solid biomass fuel combustion in local grate furnaces can be developed without a direct financial support.

In the future perspective, what could become the part of the state environmental policy contributing to the renewable energy development is either a significant increase in rates of charges collected for carbon dioxide emission (current rates are minimal), or imposing a carbon tax (such solution is recommended by some non-governmental organizations as an efficient way of adjustment to the Climate Convention's demands). Additional solution in favour of non-conventional energy producers would be to withdraw an excise tax and VAT imposed on energy generated from renewable sources. For the time being, however, these solutions are not taken into consideration by the government.

10. Institutional conditions for renewable energy development

The Ministry of Economy, which is in charge of the energy policy, pays only limited attention to issues of renewable energy development. A separate policy document concerning the non-conventional energy development strategy has been worked on for some time now by a special interministerial group established at the initiative of the Ministry of Economy. The work of the interministerial group is directed at drafting a programme of action in support of the renewable energy development. For the time being, however, according to the government's plans, renewable energy sources will not play an important role in energy consumption structure in Poland. Unlike many European governments, the Polish Ministry of Economy has not taken an obligation to ensure that within a specified period of time a quantified share in domestic energy balance will be delivered by renewable sources.

Problems related to the exploitation of renewable energy sources are regulated also by the decisions and actions of other central institutions such as e.g. Ministry of Environmental Protection (limiting emissions), Ministry of Agriculture (rural areas activization), Ministry of Finance (taxes on energy carriers, investment tax breaks, etc). For none of the aforementioned authorities is the problem of renewable sources of energy a priority or the main area of activity. Moreover, the decisions made by the bodies in question are often not coordinated with one another. Despite the claims of the environmental and pro-renewables interest groups, no government body which would be in charge of the exploitation of renewable sources of energy has been established so far (as it has been done with National Agency for Energy Conservation - an authority which supervises and supports efficiency of energy use). This is considered one of the main institutional obstacles in non-conventional energy development in Poland. Non-governmental organizations concerned with the promotion of renewable energy such as the Polish Association for Solar Energy, the Polish Association for Wind Energy, the Polish Ecological Club claim that without relevant institutions and extending of state support for producers of energy from renewable sources it will be difficult to increase the share of renewable energy in the Polish energy market.

11. Parastatal subsidies for environmentally friendly energy projects

As indicated above, there are no comprehensive integrated government programmes aimed at environmentally friendly restructuring of the Polish energy sector. Nevertheless it is possible to list programmes running, or likely to be launched in the near future, whose mission is to modernize the energy sector serving environmental protection purposes at the same time.

An example of a programme that has operated for a couple of years successfully is a competition for the best district heating project organized by the EcoFund, a foundation established (in 1992) by the Minister of Finance to administer Polish debt-for-environment swaps. The EcoFund is a non-government organization yet with an important impact on the process of environmental protection investment in Poland. Its current annual budget of 30 million USD (in 1992-1996 spent 70 million USD in total) may seem minuscule when compared with the total annual environmental investment expenditures of well over 1.5 billion USD. Nevertheless, its clearly focused priorities and expert management made it an important catalyst of change in several areas including the energy-environment interface.

In 1997 several projects concerning the competition for heating system modernizations and improvements were selected for co-financing through EcoFund. The eligibility criteria for projects were threefold:

(1) The total power installed of the system (covering an energy source, distribution and end users) is between 1 MW and 50 MW;

(2) The system must be technologically closed;

(3) The project needs to be comprehensive, i.e. addressing all three components of the system: source, distribution and users.

- Approximately 350 applicants - district heating stations, municipal governments, housing cooperatives, engineering firms etc. - requested application forms. The EcoFund received 25 eligible project proposals, of which 4 were selected for funding. The selection criteria included:

- environmental effectiveness in terms of reducing emission of three key pollutants - sulphur dioxide, nitrogen oxides and particulate matter - and carbon dioxide;
- project location, i.e. its impact on ambient environmental quality and human exposure;
- technological soundness as demonstrated by energy savings.

The EcoFund runs several competitions periodically, some of which targetted at the energy sector. They fill the niche that other government agencies have failed to take care of leading to a 'regulation failure'. Moreover the perspective of savings on energy bills proved insufficient for a great majority of energy users to undertake certain measures out of economic interest thus leading to a 'market failure'. If skillfully designed and managed, parastatal institutions like the EcoFund may compensate for both failures providing inspiration and incentives for what could have been achieved through regulation or market forces.

'Joint implementation' (JI) under the Climate Convention is another example of a mechanism of supporting energy saving measures over and above what state jurisdictions and market forces can achieve. In a JI project the government of a developed country commits certain funds to support carbon dioxide emission reductions in another (presumably less economically advanced) country leading to meeting the Convention targets jointly rather than individually. On theoretical grounds it can be argued that JI provides an environmentally sound and economically efficient mechanism of mitigating the global climate change.

Several stakeholders indicated interest in promoting JI. Unfortunately, at the same time, JI can be misused as a means to promote narrow sectoral group interests by wasting the taxpayer money on projects that lead to environmental effects definitely lower than achievable. On top of that there are unresolved problems even if projects selected are sound. The problem is caused by 'bilateralism' as opposed to arranging assistance on a multilateral basis.

Here is what typically goes wrong with bilateral assistance. (1) There is no infrastructure to select projects, or the infrastructure is replicated several times to deal with various donors thus increasing administrative costs (and the allocation of budgets planned for consultations, expert assessments etc.). (2) Projects are supply rather than demand driven. (3)

Procurement is 'tied' and hence more costly (sometimes as much as 30% above international market prices) as competition is eliminated.

It is possible to arrange and coordinate assistance on a multilateral basis by: (1) establishing multilateral facilities to select projects according to cost-effectiveness criteria; (2) balancing project budgets with respect to equipment purchases versus feasibility studies and other consulting services; and (3) 'untying' procurement. The latter point is often misunderstood and therefore requires some explanation. To eliminate tied procurement (i.e. a system where a project proposer indicates the supplier of goods and services) does not mean that the donor country loses control of what the taxpayer money is spent for. There are means to monitor the geographical distribution of contracts awarded - on a multilateral, or 'club' basis - under a given assistance scheme and periodically fine tune the procedures so that all donors receive 'fair' shares. Even though this approach is visibly superior over traditional ones, it is applied less frequently than its cost-ineffective alternatives. A well-known dictum states: "there is no constituency for efficiency". Efficient mechanisms may be prevented by narrow group interests linked to their inefficient alternatives from being adopted.

The two examples referred to in this section indicate that there exist financial mechanisms that can compensate for insufficient state regulation at the energy-environment interface. Their adequacy, however, depends strongly on project selection criteria and other operational issues.

12. Summary and conclusions

Environmental aspects of the Polish transition have been widely documented. After 1989 the pollution and resource intensity of GDP improved markedly. What can be demonstrated is that, contrary to some early expectations, environmental improvements were predating progress in energy efficiency. Pollution was reduced significantly as a result of more effective enforcement accompanied by sharply increased environmental investment expenditures.

Achieving cost-effectiveness in environmental expenditure and integration with other areas of government activities are those aspects which require substantial improvement everywhere throughout the region. For many years to come, energy sector will remain to be an important area of government concern and intervention.

As long as the full pricing of energy (including environmental externalities) is not introduced, the future of renewable energy development in Poland will depend on state intervention. The scope of such intervention should be carefully assessed. Currently, no studies indicating an optimum level of state support weight against expected benefits of more vigorous renewable energy development are available. At present situation renewables can hardly compete with conventional energy, but comparing the costs of electricity from coal, gas and various renewable technologies is complicated and potentially misleading. This is due in large pert due to the problems associated with internalising costs related to the environment and health. Furthermore, the initial state of development of many renewable technologies carries with it significant start-up costs.

The official position of the government expressed in the Energy Policy Guidelines mentions renewable energy as a means of environmental protection which should be developed. The existing legislation contains some provisions treating renewable energy in a preferential way (eg. through premium prices for certain producers), yet only to a limited degree. Project-by-project support, by means of grants and preferential loans, is provided by environmental funds. However no big scale state support programme for renewable energy development has been implemented yet. In a long run, the environmentally friendly nature of renewable energy strengthened by new technological developments and fuel diversity considerations will undoubtly lead to increased interest in tapping these resources. Though the renewables together could satisfy a large part of Poland's energy needs, economic reality alone makes it likely that coal will continue to play a role of dominant fuel, at least in the short to medium term.

With stationary (mostly industrial and municipal) sources of pollution step-by-step brought under increasingly effective control, the environmental problems in CEE will start to resemble those of developed market economies. The policy will have to focus on non-point sources of air and water pollution, including car exhaust gases. As Poland reached the number of 6 cars per 10 households already in 1995 - the level projected in the early 1990s for the year 2000 or later - this moment may come sooner than many analysts anticipate. As more and more activities are affected by environmental protection measures, cost-effectiveness considerations need to attract more attention.

References

The Economist, (1992), 'East European pollution: Dirty stories', *The Economist*, February 1, p.29.
Energy Policy Guidelines for Poland until 2010 (1995), as accepted by the Council of Ministers on October 17, 1995.
Kaderjak, P. (1997), 'Economics for Environmental Policy in the Central Eastern European Transformation: How the Context and Textbook Prescriptions Are Related?', Chapter 8, in *Controlling Pollution in transition Economies: Theories and Methods*, Cheltenham, Edward Elgar.
Ministry of Economy, Dept. of Energy and Environment (1998), 'Renewable energy'.
Ministry of Environment (1990a.), 'National Ecological Policy', Warsaw.
Ministry of Environment (1990b.), 'National Environmental Policy', Outline of Economic Instruments, Warsaw.
Ministry of Environment of the Czech Republic (1995), 'State Environmental Policy', Prague.
Ministry of Industry and Trade (MoIT) (1996), 'Renewable energy sources in state policy'.
National Communication to the Secretariat of the United Nations Framework Convention on Climate Change (1998).
Nowicki, M. (1993), *Environment in Poland. Issues and solutions*. Dordrecht, Kluwer.
Ochrona, S. (1990-1997), 'Glowny Urzad Statystyczny', in *Environmental Protection Yearbook 1990-1997*, Warsaw.
Radetzki, M. (1995), *Polish Coal and the European Energy Market Integration*, Avebury, Aldershot.
Romaniszyn, W. (1998), 'Wind energy', *Ekofinanse* nr 8/39/98.
Sayigh, A. (1997), 'Renewable energy - the way forward', in *Renewable Energy - a strategy for sustainable development*, IBMER, Warsaw.
Strategies of the GHG's emission reduction and adaptation of the Polish economy to the changed climate, *Final Report*, (1996).
Toman, M. A., Bates, R. W. and Cofala, J. (1994), 'Alternative standards and instruments for air pollution control in Poland' in *Environmental and Resource Economics* 4:401-417.
Wisniewski, G. (1998) 'Institutional and legal conditions for utilization of renewable energy in Poland', manuscript.
World Bank (1992), 'Poland. Environmental Strategy, Report No: 9808-POL', Washington DC.
World Bank (1993), 'Setting Environmental Priorities in Central and Eastern Europe'. Discussion Document on Analytical Approaches, Washington DC.
Zylicz, T. (1988), 'Will Poland Join the 30% Club' in *European Environment Review* 1/1988, pp.2-5 & 36.
Zylicz, T. (1994a), 'Environmental Policy Reform in Poland', in Thomas Sterner (ed.), *Economic Policies for Sustainable Development*, Kluwer, Dordrecht, pp.82-112.
Zylicz, T. (1994b), 'Implementing Environmental Policies in Central and Eastern Europe', in A.-M. Jansson, C. Folke, R. Costanza and M. Hammer (eds.) *Investing in Natural Capital: the Ecological Economics Approach to Sustainability*, Island Press, Washington DC, pp.408-430.

Zylicz, T. (1998) 'Environmental policy in economies in transition', in T. Tietenberg and H. Folmer (eds.), *The international yearbook of environmental and resource economics 1998-1999: a survey of current issues*, Edward Elgar, Cheltenham, pp.119-152.

Zylicz, T. and Lehoczki, Z. (1994), 'Towards Environmental Recovery. The Czech Republic - Hungary - Poland - Slovakia', in R.Domanski & E.Judge (eds), *Changes in the Regional Economy in the Period of System Transformation*, PWN, Warszawa 1994, pp.91-154.

3 The Fossil Fuel Levy: How (not) to Save Nuclear Power

PIOTR JASINSKI

1. Introduction

The transfer of assets from the public sector to the private one requires their careful valuation, and the flotation of shares in particular requires that proper accounts are produced. Privatisation of the electricity supply industry, first in England and Wales, later in Scotland and eventually in Northern Ireland, was no exception. What the splitting up of the Central Electricity Generating Board (CEGB) revealed was, among other things, the immense cost, previously hidden, of the British nuclear programme. High levels of uncertainty related to future costs and liabilities resulted in the withdrawing from the privatisation offer first of the old type Magnox nuclear power stations, and later on also the remaining ones. Depending on their location, North or South of the border, they became parts of Scottish Nuclear and Nuclear Electric respectively. Later on, *i.e.* in June 1990, the Energy Committee published one of the most damning accusations of incompetence in a Government department ever issued by a Commons select committee. Its report on the cost of nuclear power said that there was almost total failure on the part of the Department of Energy to monitor adequately cost information fed to it by the nuclear power industry.

The necessity to correct for past mistakes meant that the Government had to introduce a nuclear levy, or subsidy, officially called the fossil fuel levy, on the bills of electricity consumers in England and Wales amounting, at least initially, to well above £1 billion a year. Although the Government claimed that, "the levy did not of itself lead to an increase in the overall cost of electricity, since the costs of nuclear power incurred by the CEGB were already being paid by consumers through the bulk supply tariff before privatisation," and that, "the levy simply made these costs transparent",[1] for electricity customers in general and for energy intensive industries in particular an about 10% surcharge on their electricity bills has always been a thorn in their side - an unnecessary cost burden. It did not help that the name of this burden suggested that it was introduced with the natural environment in mind, trying to hide in this way the sad and painful reality.

Privatisation of British Energy in the Summer 1996 and the abolishing of the nuclear element of the FFL, which soon followed that IPO, provides a good opportunity to re-examine the whole issue, especially if one is interested in regulatory policy issues. The electricity supply industry (ESI) in general and the nuclear power in particular are good examples of interactions between economic and environmental regulation. Saving the virtually bankrupt nuclear power stations could in theory lead to potentially considerable benefits for the British economy, the British taxpayers (the revenue from privatisation) and for the environment, but none of these benefits was uncontroversial, and there were substantial trade-offs involved, especially from the point of view of the effect of exploiting these benefits on the coal industry. All of this had to be achieved in parallel with attempts to introduce competition to the ESI and to establish independent regulation for that sector. As first best solutions were virtually impossible to find, the whole exercise became an exercise in expedient policy making and it is precisely from this point of view that one has to look at it.

This paper - a case study of a confused policy that apparently produced quite remarkable results - will start with a re-examination of arguments in favour of nuclear power. In Part 3 the mechanism, the financial history and the consequences of the FFL will be reviewed, and the following part will concentrate of the effect that the money collected in this way had first on Nuclear Electric and then on British Energy. The remainder of the paper will provide a tentative assessment of the whole issue from the point of view of public policy, particularly from the point of view of economic and environmental regulation.

2. The rationale for nuclear power

Whatever were the arguments for the whole nuclear program in the UK, privatisation of the electricity supply industry, first in England and Wales and later on in Scotland and Northern Ireland, forced the government to examine the programme's results and to reassess its future. One had to decide whether not to close down the existing nuclear power stations, and if they were to continue their production in a new commercialised environment, a way of financing their operations and future liabilities had to be found. The solution, initially proposed, namely to bundle all nuclear power stations in England and Wales with some conventional power stations and to create a company large enough to be able to bear both the future and current costs, had to be rejected because of the pressure from potential investors. (This by the way resulted in a very concentrated generation industry, as the decision to exclude for the time being the nuclear power stations from privatisation was taken too late to change the number of companies into which the CEGB was to be divided.) And once National Power's profits, as the larger of the two new generating companies became to be known, have become directly unavailable for the purpose of supporting nuclear power stations, the money had to come from somewhere else, and the government's reluctance to use the general taxation revenues seems understandable, although not necessarily justified.

One more decision that had to be taken was whether to build any new nuclear power stations, especially that a new technology was becoming available and one power station using it, Sizewell 'B', was already under construction. This was not, however, a very difficult decision. The overall level on uncertainty about the new design and about the developments in the electricity market suggested a moratorium, and this is precisely what happened. The Trade and Industry Select Committee was to investigate the future of nuclear industry in the UK a few years later, and the report published as a result of this investigation became a major factor in the developments on which we want to concentrate our attention in this paper.

The most important arguments in favour of the preservation of the nuclear industry regarded energy security, environment, British technology. The price to be paid for all these alleged benefits was however considerable.

2.1. Energy security

The world's first industrial scale nuclear power station, Calder Hall in Cumbria, was commissioned by the UK AEA in 1956. Since 1957, eleven such gas cooled nuclear power stations, containing Magnox reactors, were built. Eight of them still function, while the remaining three were declared as having reached the end of their operational lives and are being decommissioned.

The Magnox stations were followed by a series of Advanced Gas Cooled Reactor (AGR) stations which were commissioned by the CEGB and SSEB (South of Scotland Electricity Board) between 1976 and 1988. Nuclear Electric operated five AGRs and Scottish Nuclear two. Together with the only Pressurised Water Reactor (PWR) built at Sizewell (Suffolk), which started to supply electricity to the national grid on 14 February 1995, they became the main assets of a new company called British Energy, privatised by an IPO in Summer 1996, while the Magnox reactors were transferred to a state-owned company called Magnox Electric, later merged with a publicly owned company called BNFL (British Nuclear Fuels Limited).[2]

Since the issue of energy security had been one of the most important arguments in favour of building ever more nuclear power stations, and the severely unsatisfactory results from these stations had to be dealt with in the moment of reorganisation of the British electricity supply industry, one had to ask whether energy security was still a serious problem for the UK at the time of the debate. Was nuclear energy still the best way to assure energy security?

One way to assess energy security in a given country is to look at fuel diversification, in this case in production of electricity.

Table 1 Fuels used for the production of electricity in the UK, 1980 - 1997

	1980	1990	1997
Coal	73.4	65.0	38.0
Oil	11.1	11.0	2.0
Gas	0.06	1.0	27.0
Nuclear	10.0	21.0	30.0
Hydro	0.06	1.0	1.0
Other fuels	0.01	1.0	2.0

Source: Own calculations based on: DTI, *UK Energy in Brief*, 1998, p. 18, which contains data in absolute amounts in million tonnes of oil equivalent.

The above Table 1 together with evidence from most experts seems to show that, to quote Professor David Newbery, writing in 1993, "national (energy) security has already been achieved by the current diversity of fuel supply."[3] If this was the case in 1993, fuel diversity is even less of an issue now, when Britain enjoys the widest choice of energy in its industrial history (coal, oil, gas, nuclear, and various renewables). It is only the long lead times needed to build nuclear stations that might strengthen the case for preserving a nuclear programme in anticipation of possible fuel shortages next century. At the moment, however, the pre-privatisation fears of underinvestment have not been realised, and the UK's generating capacity is likely soon vastly to exceed demand, which means among other things that in the case of a crisis there is considerable flexibility regarding the shares of individual fuels (many of the most expensive power stations are not liquidated, but only moth-balled). The fact that the UK has its own resources of coal, oil and natural gas, particularly now that these industries are all in the private sector and so are less vulnerable to trade union militancy, strengthens even further this conclusion. It is not therefore surprising that the Government's nuclear review could not 'identify any reason why the electricity market should not of its own accord provide an appropriate level of diversity.'[14]

2.2. Environment

Unlike the issue of energy security, the awareness of environmental problems is a relative latecomer to the public debate on energy policy. The main problem from this point of view regards the emissions generated when fossil fuels are burnt to produce electricity. But, since the role of the renewables is, and in the foreseeable future will be, negligable, is nuclear energy really the solution? Any answer to this question requires two factors to be taken into account: firstly, not all fossil fuels have an equally damaging effect on the environment; and, secondly, the use of nuclear energy - and even renewables - creates its own problems.

From the purely legalistic point of view, *i.e.* with regard to the various conventions and international treaties regarding, for example, CO_2, SO_2 and NO_x emissions, that the UK government has signed over the recent years, the so-called dash-for-gas has been quite providential, and helped the UK to fulfil its reduction obligations almost entirely without any additional action. In other words, partly due to the energy policy pursued previously, a considerable improvement has been possible, but it is one in which nuclear energy played a very limited role.

In addition, any environmental case which can be made for nuclear power stations has a hefty counterweight in the risks of radioactive contamination and the large amount of irradiated equipment which has to be dealt with when a nuclear station closes. The problem of nuclear waste management has already given the government a headache over the decision on whether to proceed with the Thorp reprocessing plant in Cumbria.

For all these reasons, one has to agree with one of the conclusions of the Government's nuclear review which reads as follows:

> "The Government recognises the significant contribution that the existing nuclear power stations make towards meeting the UK's current commitments regarding the limitation of emissions. The Government concludes, however, that there is at present no evidence to support the view that a new nuclear power station is needed in the near future on emission abatement grounds."[5]

2.3. British technology

The prospects of the nuclear industry would be considerably improved if two things happened: a sharp increase in fossil fuel prices, which is possible but perhaps less likely than it used to be, or an equally sharp fall in the cost of nuclear technology, which though it could occur through technological improvements in other countries (Asia, in particular, was until the current economic crisis embracing nuclear power with enthusiasm) is also unlikely. But one of the UK nuclear industry's favourite justifications for government subsidy - that it was equipping the UK with world-class expertise and technology - has largely been disproved. The industry has simply failed to fulfil its promises of delivering limitless cheap electricity or world-class technological expertise.

Although the UK led the world in developing civil nuclear power, it failed to capitalise on the achievement, and the most widespread technologies used elsewhere in the world are now non-British. The UK has lost its historic lead in developing civil nuclear technology, and therefore, to go a little further, preserving British expertise in nuclear technology should form no part of the case for nuclear power. If any future review decides that more plants are justified, generators should be freed of any obligation to buy British. Along the same lines, arguments about job creation, especially in the construction phase of a project, do not constitute a convincing argument in favour of nuclear energy.

Taking account of a combination of all these factors the Government's Nuclear Review reached the following conclusions:

> "The Government believes that nuclear power should continue to contribute to the mix of fuels used by the electricity supply industry, provided it maintains its current high standards of safety and environmental protection and is competitive."

But:

> "... against the background of the current electricity market, providing public sector support for a new nuclear power station would constitute a significant intervention in the electricity market and that current and foreseeable circumstances do not warrant such an intervention."[6]

2.4. Competition in the electricity and energy markets

However narrow our definition of the relevant market could be, electricity produced by nuclear power stations is certainly not a market in itself. Firstly, it is an important part of the UK electricity market, together with electricity produced from other fuels, like coal, gas and oil. Although this market can be further subdivided into its base, medium and peak loads, with nuclear power stations participating, in a privileged way, in the base load, ultimately all these fuels compete with each other. Thus there is no doubt that taxing electricity consumption through a levy on all electricity supplied by licensed suppliers, in order to subsidise one means of generation, for whatever reason, distorts competition. Thus it seems that the then Government, displaying another aspect of its schizophrenia, wanted to create a competitive electricity market, while subsidising one player in the market. From this point of view any support for nuclear power is distortionary. On competitive markets there is no place for uneconomic activities, and encouraging them or their continuation will always have side effects. An increase in the market share of nuclear power from 17 to 30% was doubtless, apart from the dash-for-gas, one of the factors that contributed to the demise of the British coal industry. Which also shows, by the way, how dangerous it is to subsidise or support an industry and then to change one's mind.

But this is not the only distortion created. If there was - still is? - a case for supporting nuclear energy, doing this in the way that it was eventually done additionally distorts the broader energy market, where electricity competes with other sources of energy. In this case the distortion is in favour of the other sources, as, thanks to the fossil fuel levy, prices of electricity are at least ten per cent higher than they would have been otherwise. This has also had repercussions affecting the efforts undertaken by the European Commission in order to improve the transparency of energy pricing in the European Union.[7]

3. The fossil fuel levy

Once it was decided that the existing nuclear power stations should continue their production, regardless of how convincing the arguments actually were, the government had to solve the problem of how financially to support this industry, as it was obvious that in the new competitive environment it would not survive on its own, not to mention being able to cover its future liabilities. The solution to this problem was called 'The Fossil Fuel Levy'. In this section we shall present the controversy surrounding the levy and the way in which it operated.

3.1. The controversy

The legislative structure which forms the substantive basis of the levy is set out in sections 32 and 33 of the Electricity Act 1989, and comprises three separate elements. Firstly, the nuclear component of the NFFO requires the RECs to obtain a specific amount of their generating capacity from nuclear sources. Secondly, this is supplemented by the renewables component of the NFFO, which similarly requires RECs to purchase a specified amount of their generating capacity from renewable sources. Finally, the Fossil Fuel Levy Regulations provide for the collection and distribution of a tax aimed at refunding the additional expenditure incurred in fulfilling their obligations.[8]

However, there always was considerable confusion regarding nature and objectives of the levy, partly because the government declined to clarify the issue once and for all, and partly because the situation regarding the levy has been changing almost continuously, in particular in the process of preparing the privatisation of (some) nuclear power stations. The existence of this confusion finds its confirmation in the Trade and Industry Committee's First Report: *British Energy Policy and the Market for Coal* where one can read that, "there is some confusion over the purpose of the levy." Also both in academic papers and in press reports the levy is very often discussed in an imprecise and confusing way. This stems perhaps, at least partly, from the very name of the levy: to talk about the *fossil fuel* levy gives an impression that it is applied to electricity generated from fossil fuel sources, *i.e.* coal, oil, and gas. Yet, as we have already said, one thing remains unquestionable, it was levied on all (leviable) electricity.

3.2. The mechanism

Section 32 of the Electricity Act 1989 authorised the Secretary of State (then for Energy, now for Trade and Industry), after consultation with the Director General of Electricity Supply (DGES) and the Regional Electricity Companies (RECs), to obligate each REC to obtain a certain amount of sold by them electricity from non-fossil sources.[9] The statutory requirement for consultation between the Secretary of State, the RECs and the DGES wasn to ensure that no unreasonable requirements were placed on the RECs; a default by a REC from its obligation could constitute an offence.

Given that one of the objectives of the levy was to guarantee security of supply, the RECs had to take pains to ensure that any non-fossil generating capacity they took on 'would secure' the provision of the stated capacity. That gave rise to the demanding 'will secure test' to which all prospective non-fossil capacity is subjected. It is of no relevance for the purposes of the obligation whether RECs fulfil their requirement by contracting with non-fossil generators or by owning them. However, the provision of levy funds is only permitted in respect of additional costs incurred through collective arrangements, so that collective contracting has naturally been the preferred method of securing non-fossil capacity.[10] As a result of this the RECs have established the Non-Fossil Purchasing Agency, based in Newcastle upon Tyne, which acts as their collective purchasing agent, through which all the RECs' non-fossil capacity is secured.

The nuclear component of the NFFO came into force on 31 March 1990, as part of the Electricity (Non-Fossil Fuel Sources) (England and Wales) Order 1990, pursuant to section 32 of the Electricity Act 1989. The nuclear NFFO specified the amount of each REC's generating capacity which had to be secured from nuclear sources from 1 April 1990 to 31 March 1998. The amounts totalled to approximately the entire expected capacity of Nuclear Electric during each period, thus securing demand for nuclear electricity, and thus the viability of Nuclear Electric. The nuclear obligation was divided into seven shorter periods, each with a slightly different capacity requirement for the RECs, those differences reflected the expected pattern of decommissioning, and the expected commissioning of Sizewell 'B'.

The obligation was fulfilled when the (confidential) so-called Primary Contract between Nuclear Electric and the NFPA (on behalf of the RECs) was presented to the DGES. The aim of the Primary Contract was to ensure the supply of the specified capacity to the NFPA, employing best industry practice, and in accordance with the directions of the NFPA. Under the contract, Nuclear Electric could only receive payment for electricity which was delivered, there was no payment for capacity itself. The contract also included a cap on the amount of electricity for which Nuclear Electric will receive payment from the levy. Although the payments it was receiving from the levy represented to some extent a guaranteed income, it should be noted that these payments came on top of the income that NE was receiving from the pool, *i.e.* contracts in question were the so-called contracts for differences (CfDs). This meant that NE had still to trade in the pool with other generators, which, it was hoped, was to prevent the company from developing a 'subsidy mentality'. To the same ends, the amount which Nuclear Electric was receiving from the levy was to decline each year until 1998, when it supposed to be abolished altogether.

The FFL was authorised by legislation in section 33 of the Electricity Act 1989 and the Fossil Fuel Levy Regulations 1990 as a mechanism for meeting the additional costs incurred ensuring a diversified energy base, via the NFFO. The levy is a tax which is levied on all licensed suppliers of electricity, that is RECs and second tier suppliers, and which is recoverable from their consumers in a similar way to VAT. The levy revenue is collected by the Collector (the DGES, through Coopers and Lybrand, Deloitte), whose job includes the reimbursement of RECs for the additional costs they faced in fulfilling the non-fossil fuel obligation. However, it is the responsibility of each licensed supplier, every month, to calculate the amount that he owes under the levy and to pay it to the Collector, enclosing a statement detailing his calculations.

The amount of the levy to be collected from each licensed supplier every month is calculated with reference to the total amount charged by the supplier for the leviable electricity which it supplied in the course of the relevant month. The term 'leviable electricity' can be confusing. It is defined in section 33(8) of the Electricity Act 1989 as all the electricity, *from fossil or non-fossil sources*, supplied to a licensed supplier that is engaged in securing non-fossil capacity, in relation to the NFFO, via some collective purchasing arrangement (*i.e.* the Non-Fossil Purchasing Agency). This leaves electricity generated from non-fossil sources, either by a REC itself or contracted for individually, together with any autogenerated electricity, whether generated on-site or not, (*i.e.* from a source of less than 10MW capacity) generated from autoproduction as exempt from the levy. This amounts to a very small proportion of electricity produced in the UK, not least because electricity exempt from the levy does not entitle the buyer to any reimbursement from the levy.

Section 33 (7) of the Act obliges the Secretary of State to use the powers available to him to ensure that the revenue collected from the levy is sufficient to provide each REC with its appropriate reimbursement. Should any change in the rate of the levy be necessary, the RECs must be notified at least three months in advance of the new rate's introduction. The appropriate reimbursements are calculated on a monthly basis, with payment being made for each month in which the levy is to be made. The REC receives a twelfth (*i.e.* its share) of the total advance payments made, if there are any, through the collective purchasing arrangement. It also receives a payment equal to the extra cost which it incurred in purchasing electricity through these arrangements, rather than from a fossil fuel generating station (the cost of which is taken to be the pool-output price, rather than the price level at which hedging might be available to that REC, as the strike price under a contract for differences).

The Fossil Fuel Levy Regulations 1990 came fully into force on 1 April 1990. The initial levy was set by the Secretary of State at 10.6%. For all subsequent years it was the Director General of Electricity Supply, *i.e.* Professor Stephen Littlechild, who was responsible for calculating and setting the levy rate. Changes in the levy was usually announced in December of the previous year to become effective on the following 1 April. Despite all these provisions the whole process ended up in creating arrears by December 1995, estimated around £450m. The details of the financial history of the FFL are presented in Table 2, below.

Table 2 The financial history of the fossil fuel levy, 1990 - 1997

	1990	1991	1992	1993	1994	1995	1996	1997
Rate (%)	10.6	11	11	10	10	10	3.7	2.2
Money actually paid to nuclear generators (£m)	1175	1311	1322	1166	1106			
Money paid to the renewables (£m)	0	13	26	68	99			

Source: OFFER's and Nuclear Electric's *Annual Reports*, OFFER's Press Releases, own calculations.

As one can see, the rate remained relatively stable until very recently, and the value of leviable electricity was similarly stable only in nominal terms. Perhaps the most important thing to notice is the strong upward trend in payments to the producers of renewable energy. One has also to remember that the actual amounts due to be received by Nuclear Electric for the electricity produced and supplied were unknown.

3.3. The consequences

The money collected through the FFL ultimately went to the Nuclear Electric (see Section 4, below) and to the generators producing electricity from renewable sources.[11] This were, obviously, the most important effects the levy: the money was collected and redistributed. The levy had, however, a few more consequences and characteristics, which at least have to be mentioned here and now.

First of all, the FFL has to be looked at from the point of view of its fiscal efficiency. Indeed, without such consideration it is difficult to talk about bringing efficiency into energy policy. David Newbery reminds us that efficiency requires sensible taxes as well as competitive markets, and says that, "the fossil fuel levy fails that test".[12]

But David Newbery's interpretation of the FFL is not the only one possible, and perhaps we should treat the FFL as an environmental tax? It would appear as an addition to the existing EC proposals to levy carbon based fuels (a carbon/energy tax); proposals which were originally believed to be part of the strategy to control climate change. Putting in place a regime that allows for full pricing of energy sources, their external effect included, has much to recommend it, but will a tax (a levy) not distinguishing between the effects that various fuels have on the environment be able to achieve this objective? Doubtful. Furthermore, if global warming is something to worry about, do unilateral initiatives matter? Some countries, like Denmark, Sweden or the Netherlands, think that the benefits of their own carbon taxes outweigh their costs and shortcomings. The British government, however, more or less at the same time when the levy was introduced, published a *White Paper* shelving a special tax on fuels such as coal, oil and gas which would have cut their use to improve the environment. The *Paper* made clear that a carbon tax on fossil fuels that emit carbon dioxide - the main contributor to global warming - "will not be introduced in the next few years."[13]

The FFL had also an international dimension and, because the French produce quite a lot of their electricity at nuclear power stations, some strange consequences. UK - French trade in electricity has been possible since 1961, when the first cross-channel link was completed, although the volumes traded only became really significant after 1986, with the commissioning of a 2000MW interconnector. Initially intended to provide the mechanism for mutual support in times of supply insecurity, in fact, with EdF enjoying significant surplus baseload capacity available at relatively low prices, trade has been largely unidirectional, and since privatisation the situation has become even more polarised with the UK's imports from France increasing and there being no exports at all.

Effectively, the FFL regime and its working ment that EdF could charge UK consumers up to 10% more for its electricity than it could without the levy. In 1991 its gains from these 'green ticket' payments were £95mn on 16.3TWh of electricity, suggesting a per unit gain of 0.58p/kWh. Given that this was a premium on top of a contract price which in that year was roughly 2.8p/kWh, the average price paid for French generated electricity was 3.38p/kWh, which contrasts sharply with the prices paid for electricity by French industrial users. The burden of these higher prices was passed on to the UK electricity consumers, since the non-leviable status of French electricity meant that rate of the levy as applied to leviable electricity was increased. Under the then structure of the levy, UK consumers were hit twice when purchasing French generated electricity - once by virtue of having to make payments to EdF, and again through the need for a higher levy rate to make up the shortfall in payments owed to the Levy Collector as a result of EdF's non-leviable status. Thus, paradoxically, the purchase of apparently 'cheap' electricity from EdF, under the terms of levy, was one of the causes of higher UK electricity prices. And as regards security of supply, which apart from access to 'cheap' French electricity was the second main reason for building the interconnector, the consequence of this constant supply of French electricity to the UK market has been the closure of UK pits and power stations, caused by the gradual reliance of UK distributors on French supply, which, as we have already noted, is in no way guaranteed, as witnessed by the recent industrial disputes.

The FFL was also problematic from the point of view of the EC competition law, the issue first raised immediately after the publication of the 1988 *White Paper* 'Privatising Electricity' by Tony Blair, then Labour's Energy spokesman. In February 1989 he warned that the nuclear levy could be challenged in the European Courts for breaching EC competition rules and claimed that the non-fossil fuel quota, obliging supply companies to buy a fixed percentage of power from nuclear or renewable sources, would amount in practice to an import restriction. And although the Commission did approve the levy, it also expressed its reservations regarding it, at one stage threatening to upset the privatisation of the electricity industry. This upset was narrowly averted when the European Commission did decide to give its permission for the payment of nuclear power subsidies on 28 March 1990.[14] The decision, the subject of much disagreement among the Commissioners, was finally taken on the basis that privatisation of the UK industry would help competition within the European energy market. Some Commissioners were opposed to any further subsidies being granted to the nuclear industry, while others argued against the plan on the grounds that the UK was being given more favourable treatment on state subsidies than recently accorded to other EC members.

Last but not least one has to look at the levy and its effect on electricity prices from the point of view of the UK's competitiveness. Once the brave new world of competitive electricity markets had been created it was energy intensive industries that found themselves face to face with high electricity prices. This was due, in part, to the introduction of the new charging system based around the pool, which placed the UK at odds with other European countries which continued with traditional pricing arrangements where load management was highly valued so that high load users received favourable prices. However the exact extent of the impact of more transparent contracts on fuel bills, as compared to the effect of the new design of the systems, will always remain a hotly debated issue. Nevertheless, the industry has protested loudly since the reorganisation of the British electricity supply industry, and their complaints were often published by *Financial Times*.

Some large users from the steel industry went even so far as to say that if they moved across the Channel they would pay 50% less for their electricity, which was to be the case because of high UK prices, not Continental subsidies. It was therefore no surprise, that in February 1992 a group of large customers, including Blue Circle Cement and United Engineering Steel, threatened to stop paying the levy, which represented 11% of their electricity bills, in the following April, though they did not actually follow up their threat with action.

After rising rapidly in the build up to privatisation, UK electricity prices for high load users continued to rise faster than inflation until 1993. In 1993/4 delivered prices in England and Wales for these customers were some 50% higher than in 1987/8, and the UK's comparative position relative to continental competitors had also deteriorated considerably - a situation not helped by changes in exchange rates. In the first quarter of 1994, electricity prices paid by high load users in the chemicals industry were some 20% higher in the UK than in France, whereas 2 ½ years earlier, the differential had been nearer 5%. Since the UK emerged from the ERM, the situation improved again but a recent comparison of electricity prices for selected sites in the chemicals industry in the UK and on the continent by the CIA shows that for very high load users the UK still had a disadvantage in 1995 of up to 20%.[15] And regardless of whether the introduction of the FFL increased prices or not, its existence and its height of 10%, was in this context particularly strongly resented.

4. Finances of the nuclear power stations

We have already mentioned the confusion surrounding the objectives of the fossil fuel levy. There is, however, no doubt that Nuclear Electric, having received most of the money, profited enormously from the levy. Therefore the issue that deserves our attention in this context is that of the miraculously improved performance of Nuclear Electric since its creation.

One of the most striking features of the post-privatisation history of the electricity supply industry in the UK is, as somebody put it, an almost miraculous improvement of Nuclear Electric's performance, whose market share increased from 17 to 30% and whose operating cost per unit were slashed from 5.3 p/kWh to 1.98 p/kWh. What is more, the company has began to make profit as it were 'on its own', *i.e.* without the income from the contracts signed with the Non-Fossil Fuel Purchasing Agency.

The improvement achieved consisted in keeping the inherited power stations running while at the same time reducing costs, not least by cutting down the workforce. That is why the output per employee, probably the best measure of economic efficiency, increased from 2.9 GWh per employee in 1989/90 to 6.3 GWh per employee in 1994/95, well above 100% in five years.

Table 3 Nuclear Electric's performance, 1989/90 - 1994/95

	1989	1990	1991	1992	1993	1994
NE's output (TWh)	42	45	48	55	61	59
Operating cost per unit (p)	5.3	4.5	4.2	3.7	3.2	2.7

Source: Nuclear Electric's *Annual Reports*, various years.

Table 4 British Energy's performance, 1991/92 - 1997/8

	1992	1993	1994	1995	1996	1997	1998
BE's output (TWh)	41	49.8	54.4	55.1	61.2	67.2	66.7
Operating cost per unit (p)	3.27	2.77	2.54	2.37	2.37	2.11	1.98
Output per employee (GWh)	5.0	6.3	7.3	8.0	9.4	10.6	11.7

Source: British Energy Share Offier, p. 82 and British Energy. Annual Report and Accounts 1997-1998, p. 1.

There is no doubt, as has already been stressed many times, that the improvement in Nuclear Electric's performance has been near miraculous. This is true with respect to output level and productivity as well as with respect to their financial equivalents, although the financial position could have been far better, if the industry's regulator, Professor Stephen Littlechild had not decided to interfere with pool prices. This meant that from the point of view of Nuclear Electric, prices were unexpectedly low in almost the whole period of its operation, which meant that the additional electricity produced thanks to increased reliability of its power stations was sold at a lower price.[16]

All of this would not have been possible without the money from the electricity consumers, *i.e.* without the income from the FFL, and it seems to be clear that Nuclear Electric acted upon its assumption that the nuclear premium was supposed to improve its cash flow.

5. The Nuclear Review

The debate on the support given to nuclear power in the UK culminated in the 1995 Nuclear Review and the Trade and Industry Committee's hearings and (second) report on Nuclear Privatisation of February 1996. The former resulted in the publication of the *White Paper* 'The Prospects for Nuclear Power in the UK', the conclusions of which contained the following statements about the future of the fossil fuel levy: 'All indications are that Nuclear Electric will be able to meet the full costs of meeting its liabilities when they fall due without the benefit of the levy income in the period after the privatisation of the newer stations. The Government has decided that the element of the levy paid to Nuclear Electric will cease at the same time when the AGR and PWR nuclear stations are moved into the private sector.'[17] This was consistent with the evidence presented to the TIC which almost unanimously suggested that the (nuclear element of the) levy should be abolished in the moment of privatisation of British Energy. At the same time there also appeared to be a consensus, at least among those presenting evidence to the Committee, that the renewables obligation should remain in some form, which, by the way, was not advocacy of continued existence of the levy, since the submissions recommend that renewable energy sources be supported, if at all, using finance from general taxation.

At the moment of nuclear review all of this meant that something had to be done with the Fossil Fuel Levy, and this something had to be decided about three different periods of time:

- from then until privatisation of British Energy (July 1996);
- from privatisation of British Energy until 31 March 1998, when the levy was supposed to have come to an end;
- after 1998.

As far as the period 1996-1998 was concerned, regardless of whether the FFL was supposed to improve the cash flow of Nuclear Electric or simply to cover the liabilities inherited from the CEBG, there was absolutely no reason to continue the nuclear element of the fossil fuel levy. There were however the arrears, or, to be more precise, the overdue payments for the electricity already produced and used, according to prespecified quantities and prices.[18] Of course, £450m was not a negligible sum of money and it would have been unfair to ask Nuclear Electric, or rather its successor, to forego it, but one might wonder whether Nuclear Electric's repeated declarations that it wanted the levy to be abolished in the moment of privatisation were based on the amount of money it had actually received, or rather on that it expected to receive. This in turn meant that either the levy had to continue at its then level for almost five months after British Energy's privatisation, or that the levy had to be at least 2% higher than its otherwise very low level until its end in March 1998, tentative as the exact figures might have been.[19]

What actually happened was that first, *i.e.* in December 1995 Professor Littlechild announced that the Levy would be 10% from 1 April 1996. He was able to maintain the rate at 10% because Nuclear Electric (now Magnox Electric) had agreed with the Non-Fossil Purchasing Agency (NFPA), representing the RECs, to a postponement of receipts. A reduction in the Levy later in the financial year 1996/97 and holding it at or below this new, lower level in April 1997, were, according to OFFER, dependent upon the flotation of British Energy taking place during July 1996 and upon the contract between Magnox Electric and the NFPA being amended, as presently agreed in principle by the two parties, to terminate payments for output generated after the privatisation of British Energy, and to provide the basis for the recovery during 1996/97 and 1997/98 of payments postponed from earlier years.

On 16 July 1996 Professor Littlechild did confirm that the Fossil Fuel Levy will be 3.7% for the period 1 November 1996 to 31 March 1997. Commenting this decision Tim Eggar, the then Minister for Industry and Energy, pointed out that: "Electricity bills are already at the lowest level in real terms (excluding VAT) since 1974 and consumers also received a discount of just over 50 pounds in the first quarter of this year, following the National Grid flotation. The fall in the levy rate should lead to a further reduction in electricity bills of more than 6% in a full year, or around £19 a year in a typical domestic consumers' bill".

A few month later, on 20 December 1996, Professor Littlechild announced that he would reduce the Fossil Fuel Levy to 2.2% from 1 April 1997. Professor Littlechild said: "In July 1996, I said that I hoped it would be possible to maintain or reduce the Levy from 1 April 1997. I am pleased that this has proved possible and that I am able to announce a reduction to 2.2%, a significant reduction from the present rate of 3.7%." The new rate should be sufficient to cover renewable energy commitments and the payments due to the non-privatised part of the nuclear industry. Following the flotation of British Energy, the Levy in respect of nuclear commitments will cease from 1 April 1998. This resulted in one more reduction of the Levy, this time to the level of 0.9%, effective from precisely that day.[20]

As we have already mentioned, the legal instruments introducing the Levy were to remain in force until the end of March 1998. At the same time there are contractual commitments with producers of electricity from renewable sources going well into the next century. This meant that a way of financing them had to be found rather sooner than later. An appropriate Act - Fossil Fuel Levy Act 1998 - received Royal Assent on 18 March 1998. Its main provision was to make nuclear electricity subject to the Levy.

6. British Energy: a proof of success?

The history of nuclear power generation in the UK shows that the consecutive British government do not have too much to be proud of. The whole programme lacked financial discipline from its inception, and after that state of affair had become evident during the ESI privatisation, the instruments introduced appeared to be wanting. The introduction of the Non-Fossil Fuel Obligation was followed in an *ad hoc* way by the creation of the Fossil Fuel Levy, the very name of which proved to be confusing, not to mention the contradictory pronouncements of the government and industry officials regarding its objectives. The Levy soon after its imposition became one of the most hated taxes, especially by the major energy users and the setting of its level became a victim of political expediency, which resulted in huge arrears.

On the other hand the history of Nuclear Electric and Scottish Nuclear in the years 1990-1996 and then of British Energy since 1996 looks like a success story. We have quoted above some date on the almost miraculous improvement in performance of nuclear power stations and companies owning them, and if the behaviour of share prices can be a measure of success of a privatised company, this is certainly the case with British Energy. It is true that in the foreseeable future British Energy is not going to build any new nuclear power stations in Great Britain, but it does not mean that its management will limit itself to overseeing the process of steady decline, during which process physical assets are turned into cash, paid then to the shareholders. No, since there are many (unloved!) nuclear power stations around the world, the company - the only one electricity producer in the world using exclusively nuclear power stations - wants to buy and run them. The Seven Miles Island nuclear power station - a sister of the infamous one, which almost caused a huge disaster - to be bought by British Energy together with an American partner and talks with a Canadian Company Ontario Hydro are only the first two example of implementing this new business strategy.

At the very least it seems that the companies involved made full use of the chance given to them, not such a frequent occurrence in the case of publicly owned companies, used to be bailed out for the last time until the next last time. And all of this has been achieved in the context of a competitive electricity market. Does it mean that this history of government interference with market mechanisms ended with a happy end? Even if such a conclusion would take us a little too far, there are important lessons to be learnt from this experience, the most important perhaps being that the prospect of privatisation or rather the desire of the management to have their company privatised is almost as important as privatisation itself. Perhaps more generally, the case of nuclear power in Great Britain, some aspects of which were briefly described in this paper, shows that transfer of ownership, or the prospect thereof, fully to translate into efficiency gains requires a stable, effective regulatory framework and competition. This points out another important feature of the UK's government efforts to save nuclear power, namely that these efforts - on par with promotion on renewable energy sources - were implemented with relatively little damage to the process of introducing competition in the whole of the electricity supply industry.

Nuclear power does not seem to have been a success story in other countries either. What is more, it is difficult not to suspect that in these other countries the full extent of the problems created is still to be revealed, and the actions taken by the European Commission to create a single European electricity market may only accelerate this process. From this point of view, despite all the inconsistencies and opportunism of the British policy towards nuclear power since the restructuring and privatisation of the ESI, it shows quite well how much is possible when one really has to and wants to achieve it. Whatever the official propaganda surrounding the issue and the huge cost of the rescue operation to electricity consumers, the problem seems to have been solved once and for all, and British Energy seems to be a full participant of a truly competitive although still regulated market.

The Fossil Fuel Levy belongs to the realm of economic regulation, but its name suggests - or wants to create an impression - that saving nuclear power had also an environmental aspect, and therefore the whole exercise can be looked at from the point of view of environmental regulation.

In this context, on the one hand, as already pointed out, the levy is far from fulfilling the criteria of an environmental tax. On the other hand, the increase of market share of nuclear power from 17 to 30%, only to some extent due to the commissioning of a new power station, caused a substantial reduction of emissions resulting from producing electricity from hard coal. This however, was an exercise in making the best of the situation that was not intentionally created. The reality of competitive electricity markets is such that there does seem to be any prospects for new nuclear power stations, unless their cost relative to that of other technologies falls down considerable. Nevertheless, creating a regulatory environment in which more can done for the environment with what is already available is also important. And this is yet another lesson from the Fossil Fuel Levy fudge.

Notes

1. DTI and the Scottish Office, The Prospects for Nuclear Power in the UK. Conclusions of the Government's Nuclear Review, Cm 2860, HMSO, London May 1995, p. 62.
2. *Ibid.*, p. 7.
3. David Newbery, Personal View: Fossil fuel levy fails efficiency test, *Financial Times*, 6 May 1993.
4. DTI and the Scottish Office, *The Prospects for Nuclear Power in the UK. Conclusions of the Government's Nuclear Review, op. cit.*, p. 4.
5. *Ibid.*, p. 3.
6. *Ibid.*, p. 3.
7. See Council Directive 90/377/EEC of 29 June 1990 concerning a Community procedure to improve the transparency of gas and electricity prices charged to industrial end users, O.J. L 185, 17.07.1990 and the December 1995 *White Paper* "An energy policy for the European Union" (COM (95) 682).
8. For a more comprehensive description of the technical details see: Andrew Grenville: 'Fostering the Non-Fossil: Control of Fuel Sources for Electricity Generation in the United Kingdom, '*Journal of Energy and Natural Resources Law*', vol. 10 (1992), no. 3, p. 285 - 292.
9. Indeed the Fossil Fuel Levy is a product of the Non-Fossil Fuel Obligation. In this sense, it is possible to see the levy very much as an *ad hoc* solution to what were seen as the problems stemming from the NFFO, specifically, the increased costs faced by firms in fulfilling their obligation. This explains our considerations of factors relating to the NFFO generally, and the renewables obligation within the framework of this paper as well as in the one written for this project by Cathryn Ross.
10. See, A. Grenville, *op. cit.*, specifically p. 286. See also, Electricity Act 1989 Section 33 (8).
11. See the paper prepared for this project by Cathryn Ross.
12. David Newbery, *op. cit.*

[13] Quoted in: John Hunt, Environment White Paper: Blueprint for fossil fuel charges postponed - Taxes, *Financial Times*, 26 September 1990.
[14] *Financial Times,* report of 29 March 1990.
[15] Figures from internal research by the Chemical Industries Association.
[16] The effects of the price cap on Nuclear Electric were quite tangible. Two years later, when announcing the results for 1994/95 Dr Robert Hawley, the company Chief Executive said that if the OFFER pool price caps had not been imposed - and the time-weighted price remained at the previous year's level - NE would have made a healthy profit of around £150m.
[17] DTI and the Scottish Office, *The Prospects for Nuclear Power in the UK, op. cit.*, p. 5.
[18] In December 1995 Professor Littlechild disclosed that these payments were (then?) equal to about £450m. This is equal to 40.6% of the money paid to the nuclear generators in 1994/95.
[19] This was not, however, the only solution, and, at least in principle, it will be possible for the Government to pick up the bill for the arrears. The argument for its doing so is particularly strong if we accept that the levy was supposed to cover Nuclear Electric's liabilities and was spent on other things, at the same time helping NE to achieve the spectacular improvement of its financial results. Further more, it is the Government, of which the DGES is a part ('a non-ministerial government department'), that is responsible for the arrears in the first place.
[20] Offer's Press Release of 19 December 1997.

4 The Promotion of Renewable Energy in England and Wales: the Use of the Non-Fossil Fuel Obligation

CATHRYN ROSS

1. Introduction

This paper aims to review the use of the system known as the non-fossil fuel obligation in the accorded of renewable energy sources in England and Wales in recent years, and to evaluate the results of its use. In order to undertake such as assessment it is necessary to begin by considering the aims of renewables support in England and Wales, since once these are understood in will be possible to go to assess the success of policy in achieving them and perhaps to conclude that the most serious problems lie in the aims themselves. We will then move on to outline the rules set down and the instruments used for the support of renewables in England and Wales, before giving extensive consideration to the procedures and outcomes of the Fossil Fuel Levy and the Non-Fossil Obligation. Arrangements in Scotland are quite different and are not considered here. Particular attention will be paid to the problems experienced in the functioning of the NFFO-FFL system, and, finally, to the consideration of possible future options for renewables support.

2. An overview of renewable support in Britain

Before beginning our detailed discussion of the use of the non-fossil fuel obligation as an instrument for the support of renewable energy in England and Wales, it is useful to spend a little time placing the NFFO in the historical and wider policy contexts. In this section then we will try to show how the different non-fossil fuel obligations fit together, how they are funded and how the NFFO system fits with other support for renewables through the government's research development and demonstration (RD&D) programme.

2.1. Research development and demonstration support

The UK government's support for renewable energy began in the mid-1970s in the wake of the OPEC oil shocks, with the creation of an RD&D programme dedicated to renewable technology. The first comprehensive attempt to delineate a policy on renewables took place in 1982 when the government Advisory Council on Research and Development for fuel and power (ACORD) undertook a review establishing an approach for the appraisal of particular renewable technologies so that decisions could be taken to support them or not, based on their technological and crucially also their economic potential. This methodology was reviewed in 1986, and there have been two subsequent reviews of renewable energy support in 1988 and 1994. Throughout these reviews there has been no significant change in the approach to renewables support - a gradual decline in government support is envisaged and an increase in support from industrial and other sources in expected. The latest review has confirmed that the support programme will continue until 2005 (pending a further review in 2000).

Within this RD&D programme, the largest share of expenditure has been on wave power, followed by wind, geothermal and solar power. However, the percentage of spending on wave power fell dramatically between 1978 and 1991 from 60% to just 10%, while the percentage of funds allocated to wind power increase over the same period to around 30-40%, and by 1994 wave power had been classed in such a way as to make further research funding extremely unlikely. Support for geothermal energy peaked between 1981 and 1984 at around 30-40% and has since declined. The popularity of biomass and photovoltaic programmes have been steadily increasing, while funding for tidal and solar power has remained at a low level throughout. These changes in funding priority, some analysts have argued, have produced a lack of continuity which has significantly damaged progress in this area (Mitchell, 1996).

The National Audit Office recently undertook an inquiry in to the RD&D support offered by the Department of the Environment, aiming to evaluate its effectiveness in meeting its goals. In particular, the inquiry centred around the support offered to landfill, hot dry rocks (HDR) and wind technologies. It published its results in 1994 (NAO, 1994), and concluded that although the Department's methodology was sound, there had simply been insufficient support for projects aimed at the export market. The report also noted that the CEGB, as the main customer for energy sources, had influenced the research programme so that technologies in which it was particularly interested received disproportionate support. Thus large amounts of support were given to large scale wind power, HDR and tidal power - often to large scale and expensive projects which failed to succeed.

Before the introduction of the wsystem, then, support for renewable technologies in the UK was extremely limited, with the main renewable technologies in use being hydroelectric stations in Scotland and pumped storage in Wales. There were around 200 hydro-storage plants in the UK, but most were smaller than 300kW and were often financed not by government but by private individuals. The use of sewage gas for electricity generation was being explored by water companies in line with the aim - within the privatisation process - to reduce sewage discharge into the sea. There were around 20 small anaerobic biogas plants, supported either by small companies or private individuals. There were 32 landfill sites, all financed by the companies operating them, and mostly used for heat, rather than electricity, production. Just four municipal and general waste massburn incinerators existed - in Edmonton, Nottingham, Sheffield, and Coventry. There was also private investment in wind power consisting of six single turbines of 65-95kW dnc, several turbines of 5kW and several 25-500W turbines, as well as the ten government supported turbines between 130 and 3000kW installed at six sites throughout the country. Prior to the NFFO, all these facilities, if connected, were able to sell their electricity to the grid under the terms of the Energy Act 1983. Under the same Act electricity boards were obliged to purchase independently generated electricity, but on average the renewable generators were paid 30% *less* than the CEGB (Mitchell, 1996).

The NFFO system, then, did not constitute the starting point for renewables support in England and Wales. Rather it should be noted that the NFFO and FFL were only two (related) policy instruments within a wider government policy on renewable energy, and that this policy also belongs in the appropriate historical context.

2.2. The Non-Fossil Fuel Obligation

The Non-Fossil Fuel Obligation is a product of section 32 of the Electricity Act 1989, which authorises the Secretary of State (then for Energy, now, owing to government restructuring, for Trade and Industry), after consultation with the Director General of Electricity Supply (DGES, the head of the Office of Electricity Supply) and the public electricity suppliers (PESs), to obligate each PES to obtain a certain amount of their generating capacity from non-fossil sources. The PES (effectively the RECs)[1] pay a premium price for this non-fossil electricity and the difference between this premium price and the average monthly pool purchasing price is then reimbursed to the REC by the Non-Fossil Purchasing Agency using funds collected from the Fossil Fuel Levy. The statutory requirement for consultation between the Secretary of State, the RECs and the DGES should ensure that no unreasonable requirements are placed on the RECs; a default by a REC from its obligation can constitute an offence.

Given the need for security of supply, the RECs must take pains to ensure that any non-fossil generating capacity they take on 'will secure' the provision of the stated capacity. This gives rise to the demanding 'will secure test' to which all prospective non-fossil capacity is subjected. It is of no relevance for the purposes of the obligation whether RECs fulfil their requirement by owning or by contracting with non-fossil generators. However, the provision of levy funds is only permitted in respect of additional costs incurred through collective arrangements, so that collective contracting has naturally been the preferred method of securing non-fossil capacity.[2] As a result of this the RECs established the Non-Fossil Purchasing Agency (NFPA) which acts as their collective purchasing agent, through which all the RECs' non-fossil capacity is secured.

To be more specific, from its inception in March 1990 until the end of March 1998, the NFFO comprises two separate obligations. The most onerous obligation has been that which compels electricity suppliers to purchase a certain percentage of their electricity from nuclear sources - the 'nuclear obligation' (see Jasinski, 1997). The less onerous, though more interesting from our point of view, is that which compels suppliers to purchase a certain percentage of their electricity from renewable sources - the 'renewables obligation'. They serve effectively to provide nuclear and renewable energy producers, respectively, with a guaranteed contract for supply.

2.2.1. The Nuclear obligation

While this paper is most concerned with the effects of the NFFO and the Levy on the renewable energy, since support for nuclear and renewables share the same policy instrument, and since the lion's share of Levy funds have consistently been awarded to the nuclear industry, it is necessary briefly to consider the nuclear element of the Non-Fossil Fuel Obligation.

The NFFO, including the nuclear component, came into force on 31 March 1990, as part of the Electricity (Non-Fossil Fuel Sources) (England and Wales) Order 1990, pursuant to section 32 of the Electricity Act 1989. The nuclear NFFO specifies the amount of each REC's generating capacity which must be secured from nuclear sources from 1 April 1990 to 31 March 1998. The amounts total to approximately the entire expected capacity of the nuclear industry in England and Wales during each period, thus securing demand and therefore the viability of the industry. The nuclear obligation is divided into seven shorter periods, each with a slightly different capacity requirement for the RECs, these differences reflect the expected pattern of decommissioning, and the expected commissioning of Sizewell 'B'. The obligation is fulfilled when the (confidential) so-called Primary Contract between the nuclear electricity producers and the NFPA (on behalf of the RECs) is presented to the DGES. The aim of the Primary Contract is to ensure the supply of the specified capacity to the NFPA, employing best industry practice, and in accordance with the directions of the NFPA. Under the contract, nuclear producers may only receive payment for electricity which is delivered, there is no payment for capacity itself. The contract also includes a cap on the amount of electricity for which the English and Welsh nuclear power industry will receive payment from the levy. Although the payments it receives from the levy represent to some extent a guaranteed income, it should be noted that these payments come on top of the income that English and Welsh nuclear producers receives from the pool, *i.e.* contracts in question are the so-called contracts for differences (CfDs). This means that nuclear producers must still trade in the pool with other generators, which, it is hoped, should prevent the company from developing a 'subsidy mentality'. To the same ends, the amount which the industry receives from the levy declines each year until the end of March1998, when it should be abolished altogether. At the time of writing it is not yet known precisely how the cessation of the levy will affect financing of renewable support in England and Wales.

2.2.2. The Renewables Obligation

In a similar way to the nuclear obligation, the RECs are also obligated to secure a specified amount of capacity each year from renewable sources, such as wind, solar, tidal, wave, geothermal and biogas. At the time of writing four renewables Orders have been completed (1990, 1991, 1994, 1996) (see below), and the fifth has been announced.

The RECs' renewables obligation is measured in terms of 'declared net capacity' (dnc), which must be secured. Normally dnc is defined as the maximum level of generation which can be maintained by a plant indefinitely, without causing damage to that plant. However, given the more tenuous nature of some renewable power sources, the definition has been amended. The actual required dnc in this case becomes the normal dnc multiplied by 0.17 for solar power, 0.43 for wind power, or 0.33 for tidal or wave power. No allowance is made for stations powered by any other type of water power. Since many of the projects being assessed by the NFPA, on behalf of the RECs, only have available statistics relating to planned net capacity, a fairly stringent test is applied to ensure that only the projects with a very high probability of producing the expected capacity will receive contracts. To pass the 'will secure test' a project should:

- be able to demonstrate that there are good prospects for obtaining the necessary consents for electrical connection;
- have a defined site available, preferably with planing permission, or a good chance of getting planing permission in the near future;
- be able to secure arrangements for the supply of raw materials (fuels) and disposal of waste products, as required;
- be a technically viable scheme, with a realistic projected output;
- clearly be able to be operational on or before the contracted commissioning date;
- be able to justify the projected capital and operating costs of the scheme;
- be able to demonstrate that adequate funding is available.

In submitting a bid to supply part of a REC's renewables obligation, a generator must obtain a tender pack from the NFPA, which will require him to provide such information as to equip the DGES to conduct the 'will secure test'.

In NFFO-2 'technology bands' were introduced so that wind power competes with wind power, biomass with biomass, tidal power with tidal power, and the band system continues today (see below). The contract which a successful generator must sign may take one of two forms. The first concerns non-pooled schemes, which provide less than 10MW of capacity to a particular REC's distribution system. The second applies to schemes providing more than 10MW of capacity, and which are, necessarily, pooled. Whichever contract applies, the generator is paid for delivered electricity only (as in the nuclear obligation), with no payment being made for capacity. The contracts involved here are contracts for differences. The strike price is the same for every contract within a given technology band and is decided by the NFPA through a bidding process in which each generator must submit the price for which it is prepared to accept a contract with the NFPA. Following this the NFPA accepts the tenders in ascending order of price until it has fulfilled its requirement. The strike price of the last accepted bid, *i.e.* the highest accepted price, is adopted as the price paid to all generators within that technology band. Although the contracts were intended to involve payment to the generators from the fossil fuel levy the form they take ensures that payments are, in practice, independent of the receipt by the NFPA of any revenue from the levy.[3]

2.3.The Fossil Fuel Levy

The fossil fuel levy was established by legislation in section 33 of the Electricity Act 1989 and the Fossil Fuel Levy Regulations 1990 as a mechanism for meeting the additional costs incurred ensuring a diversified energy base, via the NFFO. The levy is a tax which is levied on all licensed suppliers of electricity, that is RECs and second tier suppliers, and which is recoverable from their consumers in a similar way to VAT. The levy revenue is collected by the Collector (the DGES, through Coopers and Lybrand, Deloitte), whose job includes the reimbursement of RECs for the additional costs they faced in fulfilling the non-fossil fuel obligation. However, it is the responsibility of each licensed supplier, every month, to calculate the amount that he owes under the levy and to pay it to the Collector, enclosing a statement detailing his calculations.

The amount of the levy to be collected from each licensed supplier every month is calculated with reference to the total amount charged by the supplier for the leviable electricity which it supplied in the course of the relevant month. The term 'leviable electricity' can be confusing. It is defined in section 33 (8) of the Electricity Act 1989 as all the electricity, *from fossil or non-fossil sources*, supplied to a licensed supplier that is engaged in securing non-fossil capacity, in relation to the NFFO, via some collective purchasing arrangement (*i.e.* the Non-Fossil Purchasing Agency). This leaves electricity generated from non-fossil sources, either by a REC itself or contracted for individually, together with any autogenerated electricity, whether generated on-site or not (*i.e.* from a source of less than 10MW capacity) generated from autoproduction as exempt from the levy. This amounts to a very small proportion of electricity produced in the UK, not least because electricity exempt from the levy does not entitle the buyer to any reimbursement from the levy.

Section 33 (7) of the Act obliges the Secretary of State to use the powers available to him to ensure that the revenue collected from the levy is sufficient to provide each REC with its appropriate reimbursement. Should any change in the rate of the levy be necessary, the RECs must be notified at least three months in advance of the new rate's introduction. The appropriate reimbursements are calculated on a monthly basis, with payment being made for each month in which the levy is to be made. The REC receives a twelfth (*i.e.* its share) of the total advance payments made, if there are any, through the collective purchasing arrangement. It also receives a payment equal to the extra cost which it incurred in purchasing electricity through these arrangements, rather than from a fossil fuel generating station (the cost of which is taken to be the pool-output price, rather than the price level at which hedging might be available to that REC, as the strike price under a contract for differences).

It is also worth noting that, as an industrial support measure, the terms of the fossil fuel levy had to be negotiated with the European Commission. Initially plans had been submitted to the Commission involving support for non-fossil fuels - pointedly no mention of nuclear power explicitly - from the fossil fuel levy over an indefinite period, likely to be at least fifteen years. However, the Commission was unwilling to accept indefinite support said instead that it "in 1990 did not wish to grant authorisation for support of nuclear power beyond 1998" (HoC, 1992, p. 15:5) not least because of the hindering effect the levy was likely to have on the introduction of competition into this sector. Thus, the fossil fuel levy must end in March 1998 and it is not yet known what effect this will have on the financing of support for renewable energy sources in England and Wales.

2.4. *The aims of renewables support under the NFFO in England and Wales*

There can be no doubt (see Jasinski, 1997) that the primary aim of the FFL-NFFO system was not the support and encouragement of renewable energy in the UK. Indeed, 'the renewables NFFO developed out of the need to find a means of supporting nuclear power, once it was realised that the nuclear portion of the electricity supply industry could not be privatised in 1989,' (Mitchell, 1996, p. 169). That noted, however, there has been an element of recent government energy policy that has sought to contribute to the use and development of renewables, so that, in spite of the (deliberate?) confusion over the true aims of the FFL-NFFO system, it is viable to discuss the aims of the government's renewables support policy.

Perhaps the clearest statement of policy towards renewable energy in the UK came from the Minister for Energy (within the Department of Trade and Industry) in July 1993 (DTI, 1994), when he said that:

> "Government policy is to stimulate the development of new renewable energy technologies where they have the prospect of being economically attractive and environmentally acceptable in order to contribute to:
> - diverse secure and sustainable energy supplies;
> - reduction in the emission of pollutants;
> - encouragement of internationally competitive renewable industries."

explaining further that:

> "the purpose of the NFFO Orders is to create an initial market so that in the not too distant future the most promising renewable can compete without financial support. This requires a steady convergence under successive Orders between the price paid under the NFFO market price. This will only be achieved if there is competition in the allocation of NFFO contracts."

In keeping with the philosophy of the consecutive Conservative governments of the last decade, it was consistently emphasised in publications connected to renewables policy - and in the above quote - that the government saw its role rather as to enable the market to support renewable energy than to support it directly itself. Indeed, a 1990 report[4] made quite clear the fundamental objections of the government to any kind of tax to support environmental improvement. Such an idea was only considered as an addendum to a nuclear tax, probably in order to placate public (and European Commission) opinion in this politically sensitive area. With the announcement of the Fourth Order in 1996, the emphasis shifted even more towards a market approach with respect to renewables support, there was no mention of the environmental aims of the NFFO, simply a statement that the Order is intended to encourage convergence between electricity prices under the Order and market prices for electricity.

At least partly because of this governmental attitude, and the reluctance to subsidise only those technologies that were very likely to stand on their own in the future, the government took great care to discriminate between different types of renewable energy technology. Types of renewable energy that were likely to be most fruitful for the UK were: fuel cells, energy crops, photovoltaics, hydroelectric power, wastes, onshore wind, and solar. Wave, geothermal, tidal and offshore wind power have been classified as unlikely to be fruitfully developed in the UK and have been deprived of funding on these grounds.

As Mitchell (1996) points out, the initial objective of the Conservative government with regard to renewable power was the achievement of just 600MW dnc by 2000, which Friends of the Earth described as a, 'pathetically low and unambitious target [which] sums up the Government's attitude to the future of renewables.'[5] This was later revised to 1000MW and finally to a capacity of just 1500MW dnc or just 3% contribution of renewables to total electricity consumption by 2000.[6] In 1991 Greenpeace launched a campaign to increase the target to 10% of total consumption. And it seems that all this has had some effect, since the current Labour government has revised the target to just that level. It is now the aim of government to ensure that 10% of UK electricity consumption is accounted for by renewable generation by 2010. The exact amount of dnc required to fulfil this, though, is necessarily unclear - if electricity consumption were to remain at 1995 levels 10% of consumption would mean 5100MW of dnc, although if consumption increases as is estimated it may mean up to 6400MW dnc (IIEC, 1997, p. 31). While this revision has been generally greeted favourably, it should be noted that a 1991 study by the Energy Technology Support Unit (Harwell) concluded that it would be possible to have 25,000 MW renewable capacity in the UK by 2010, so that even the current targets might be considered relatively unambitious.

3. A note on the Fossil Fuel Levy as *Ersatz* Carbon Tax

As we have already discussed, the events and motives surrounding the origins of the fossil fuel levy are somewhat unclear although among the various justifications for the levy the idea that it is in someway an 'environmental tax' or more specifically some type of carbon tax recurs. Thus, it is worthwhile to spend a little time in discussing the extent to which the FFL measures up in these respects.

On the most general level, it is true that the FFL has had positive environmental effects. One of the most significant effects of the levy has been to increase the price of electricity over and above what it would otherwise have been (see Jasinski, 1997). Assuming some that electricity is a normal good, whatever is the magnitude of its elasticity, this increase in price should have resulted in lower levels of consumption than there would have been without the levy. Even if such lower levels are not apparent in the short run owing to inelastic demand for electricity, the higher price should serve to make other forms of power more viable and to stimulate innovation in energy saving techniques.

A further result of the existence of the FFL-NFFO system has been the support of nuclear power, which in itself has environmental repercussions. It could be, and has been, argued that the support provided by the Levy to nuclear power has environmental advantages, since nuclear power production involves the release of substantially less CO_2 than convention fossil fuel stations. And nuclear power is also to some extent 'renewable' via reprocessing. However, there are also environmental problems with nuclear power that, at various times in recent history, have been all too obvious. The risk of leakage and radioactive contamination[7] is a very real one and the enormity of the damage that even a moderate scale accident could bring demand that anyone with an appreciation of the precautionary principle take a rather dim view of nuclear power. And even if, during the lifetime of the nuclear station everything functions according to plan, there is always an inevitable problem with decommissioning at the end of a reactor's life, a costly and difficult problem where experience is very limited.

More specifically, beyond these broad environmental effects, there should be no mistaking the fact that the fossil fuel levy is simply *not* a carbon tax. Admittedly, there may be some superficial similarities, which have been exploited by those wishing to make the levy more acceptable, but on closer examination it quickly becomes apparent that there exist substantive differences between the levy and any form of carbon tax.

Firstly, the fossil fuel levy is levied on *all* electricity supplied by licensed suppliers - including that from renewable sources. A tax which genuinely sought to discourage the use of fossil fuels as part of an overall environmental strategy, would simply apply a levy to all electricity produced from fossil fuel sources, and on no other electricity.

Secondly, the base unit of the levy is not the amount of CO_2 (or for that matter any other hydrocarbon) involved in the generation process but rather the amount of electricity produced, regardless of the emissions created in its generation. In the UK, this has resulted in a situation where electricity from combined cycle gas turbine (CCGT) power stations attracts the same levy per unit as electricity generated using coal and oil, which produces relatively higher carbon dioxide emission levels. A genuine carbon tax would be calculated in relation to the amount of carbon dioxide emitted in the course of electricity generation, and indeed would probably form part of a wider scheme of more pro-ecological taxation, not existing in the UK, which applied taxes to discourage the use of all fuels generating carbon dioxide, including, for example, petrol in cars.

The only prospects for a genuine carbon tax in the UK look to depend on Europe. The Community has been toying with the idea of a carbon tax for many years, and it is an idea which has met with general approval from the Member States (excluding the UK). However, as yet there is no agreement on the precise form that such a tax should take, let alone on any date for its implementation. And the sheer contention surrounding the award of any fiscal powers to the European level, combined with the immense strength of the industrial lobby should mitigate against any progress in the near future.

4. Renewables support under the NFFO-FFL system in practice

Following the previous overview of the instruments of government policy to support renewables in England and Wales, this section will attempt to examine just what the NFFO-FFL instrument in particular managed to achieve, and where the problems with it lie. In particular, attention will be paid to the functioning of the successive Orders, and to what results they achieved.

The NFFO-1

The first non-fossil fuel obligation Order was issued in 1990 and contracted for just 102MW dnc of new capacity from renewable sources. This contracted capacity was supplied from various sources (see Figure 4.1 below), almost one quarter coming from municipal and general industrial waste. Hydro and wind sources both accounted for around 8% of contracted capacity, with sewage gas accounting for around 4.2%.

Given the rather murky origins of the renewables obligations, it can hardly be surprising that the procedures involved in NFFO-1, involved a certain amount of confusion as to the allocation of responsibilities between the RECs, OFFER and the NFPA - all recently established. NFFO-1 relied on a process of 'cost-justification' by the applicant, individually judged by the RECs with no direct competition between different projects or different technologies. The degree to which individual RECs were prepared to push the applicants in justifying their costs varied significantly. In addition to the variation in assessment, the details of the process employed also varied through the course of the competition so that applicants were not always entirely certain of what was required of them. Indeed, assessing authorities also seemed at times confused. OFFER, for example, very strictly enforced the 'will-secure test', assessing projects on the basis of financial performance, site availability and planning permission, technical viability, feasibility of the commissioning date, estimations of capital and operating costs, and availability of funds.

One result of their rigour was that wind power projects, because of the interruptibility of their supply, were very harshly treated. Wind power also suffered particularly from the introduction by the NFPA (on the advice of the Department of the Environment) of a maximum price of 6p/kWh more than which the Agency declared itself unwilling to pay for renewable power. Such was the effect on wind power that the Agency later raised its price ceiling to 9p/kWh for some wind generators (Mitchell, 1996, p.172). The whole exercise was viewed rather dimly in many quarters. Friends of the Earth were moved to comment that, 'the government has squeezed out viable projects by continually changing the rules.'[18]

Table 1 Summary of the 1990 Renewables Order (NFFO-1) and Associated Contracts

Technology Band*	No. of Schemes Contracted	No. of Schemes Commissioned as at 30/9/97	Contracted Capacity (MW dnc)	Capacity Commissioned at 30/9/97 (MW dnc)	Highest Price Paid (p/kWh)
Hydro	26	21	11.85	10.00	7.50
Landfill Gas	25	19	35.50	30.78	6.40
Municipal & General Industrial Waste	4	4	40.63	40.63	6.00
Other	4	4	45.48	45.48	6.00
Sewage Gas	7	6	6.44	5.98	6.00
Wind	9	7	12.21	11.70	10.00
Totals	75	63	152.11	144.57	-

Source: Table 1, *Renewable Energy Bulletin* No. 7, 26 November 1997, Annex G.
NFFO-1 was not divided into technology bands, though it has been presented in this way here for the purpose of comparison.

The NFFO-2

The most obvious change between NFFO-1 and NFFO-2 is that while NFFO-1 contracted for just 102MW on new capacity (dnc), NFFO-2 imposed an obligation that built up 457MW dnc, an increase of nearly 350% in contracted capacity as compared to NFFO-1. Within that contracted capacity, as Figure 4.2 (below) shows, municipal and general industrial waste still dominated and had increased its share to 58% of the total, in absolute terms an increase of 40.63MW dnc to 271.48MW dnc. The share of landfill gas had fallen from around 23% in NFFO-1 to around 10% in NFFO-2, although in absolute terms the contracted capacity from landfill gas actually rose from 11.85 MW dnc to 10.86 MW dnc. Similarly, while the share of sewage gas in contracted capacity increased only slightly from around 4% to approximately 6%, the actual amounts contracted for increase substantially from 6.44MW dnc to 26.86 MW dnc. The share accounted from by wind power - still undivided at this point - increased from 8% in NFFO-1 to around 18% in NFFO, with absolute amounts increasing from 12.21MW dnc to 84.43MW dnc. By contrast, the amount of capacity contracted for from hydro remained more or less the same at around 11MW dnc, and so correspondingly, its share of total contracted capacity declined dramatically from almost 8% in NFFO-1 to just over 2% in NFFO-2.

A further major change between the system used in NFFO-1 and that used from 1991 in NFFO-2 - and currently - was the establishment a competitive bidding process which takes place within different technology bands. Under this system each project must submit a bid price, and a strike price is set for each technology band. All projects which submit a bid price below the strike price are then offered contracts at the strike price. Projects applying for funds from the levy may apply for finance in the form of grants, loans (at low interest rates) or a combination of the two. All projects must pay off their capital borrowing by the time their guaranteed NFFO contract expires. Also by the announcement of NFFO-2, further problems with the NFFO-FFL system had become apparent. Small scale projects and those from independent developers were finding the competition for projects very tough and the odds in many ways stacked against them. All but two of the independent projects in NFFO-2 were forced to take equity from RECs, generators or water companies at a high cost of capital, thus leading lower returns (Mitchell, 1996, p. 172). Further, the 1998 end-date for the NFFO-FFL lent a degree of urgency to renewable projects that was problematic in the larger scale, more complex projects. In the case of wind energy the haste inspired by the end-date lead contract holders to buy foreign turbines and in fact of the 1990 and 1991 wind projects 83% of turbines were imported (Windirections XIII (4) 1994, p. 13) leading the House of Commons Select Committee on Welsh Affairs[9] to say, 'it is doubtful that another mechanism could have been more successful in supporting a foreign industry than compelling all developments to occur within a short period of time when the domestic industry is in its infancy and anyway tied to one developer.' Further, more than 200MW dnc of waste-to-energy projects had to be terminated because of fears that the 1998 end-date allowed insufficient time to develop them satisfactorily (Mitchell, 1996, p. 173).

Table 2   **Summary of the 1991 Renewables Order (NFFO-2) and Associated Contracts**

Technology Bands	No. of Schemes Contracted	No. of Schemes Commissioned as at 30/9/97	Contracted capacity(MW dnc)	Capacity Commissioned at 30/9/97(MW dnc)	Band Strike Price(p/kWh)
Hydro	12	10	10.86	10.46	6.00
Landfill Gas	28	26	48.45	46.39	5.70
Municipal & General Industrial Waste	10	2	271.48	31.50	6.55
Other	4	1	30.15	12.50	5.90
Sewage Gas	19	19	26.86	26.86	5.90
Wind	49	25	84.43	53.82	11.00
Totals	122	83	472.23	181.53	-

Source: Table 2, *Renewable Energy Bulletin* No. 7, 25 November 1997, Annex G.

The NFFO-3

In terms of contracted capacity, NFFO-3 saw a further increase - of 170MW dnc - over NFFO-2 with 627MW dnc contracted for in this Order. As Table 4.4 shows (when compared with Table 4.2) the structure of the technology bands was altered with the sewage gas band abandoned, and energy crops and agricultural and forestry waste (ECAFW) included. Significantly the wind technology band was divided into less than 1.6MW dnc and greater than 1.6 MW dnc, to counter the criticism that under the undivided band small wind generation projects had been losing out. The total share in contracted capacity accounted for by wind power did increase from approximately 18% in NFFO-2 to just over 23% in NFFO-2, an increase of just over 81MW dnc to 165.6 MW dnc. Other new technology bands in NFFO-3 were for energy crops and agricultural and forestry waste (ECAFW), divided into that used for gasification and that not (residual). In total ECAFW accounted for 19.9% of contracted capacity, with the majority of that (16.6% of total) coming from the residual ECAFW.

    Other changes in the structure of contracted capacity were evident. The share of landfill gas increased slightly to around 13% of total contracted capacity, representing an increase of 33.55MW dnc to 82MW dnc. The share of hydro remained exactly the same at 2.3%, although in absolute terms contracted capacity from hydro increased by around 3.5MW dnc from NFFO-2 to 14.4MW dnc. Contracted capacity from municipal and general industrial waste sources still dominated the NFFO-3, although its share in the total had fallen to almost 39% and it accounted for nearly 30MW dnc less in this Order than in the previous one, with contracts for 242MW dnc.

Although the third tranche award of 1994 saw the projects judged in relation to a time period of twenty years, which was certainly an improvement on the eight year period allowed in the first tranche, some argue that it is still too short to allow proper consideration of some worthwhile projects. Although the standard period over which conventional power station viability is judged is just fifteen years, renewables projects, they argue, require a more long term view. Developments which might be best in both environmental and economic terms when judged over a twenty or twenty five year period, can appear infeasible over a fifteen year timespan. At least partly in recognition of the fact that the extension of the contract period to 15 years was reducing fixed costs, in NFFO-3 changes the price basis for renewable funding.

Another change was made to the prices received by renewables contractors. Previously, all renewable generators which received contracts were paid at the strike price regardless of what their bid price had been, in this tranche, however, all contract holders were awarded their bid price. Table 4.3, below, shows just what difference the abandonment of strike price payment in NFFO-3 had on prices per kWh.

The effects of these price reductions certainly left renewable generators with less room for manoeuvre. However, it should be stressed that the increasing experience with renewables generation in England and Wales over time mean that prices for equipment involved in generation have fallen, and the prices charged by planners, lawyers and other related professionals have also come down. There have also been technical improvements in the generation process making it more efficient and bringing down the cost per kWh. Costs for wind power especially fell, from £1000 per kWh in NFFO2 to £700-£750 per kWh in NFFO-3 (Mitchell, 1996, p.175).

Table 3   Prices under the NFFO-2 Strike Price System and NFFO-3 Bid Price System

Technology	Band Price NFFO-2 price/kWh	Band Price NFFO-3 price/kWh (average)
Wind	11.0	4.32 (1.6 MW dnc or over) 5.29 (under 1.6MW dnc)
Hydro	6.0	4.46
Landfill Gas	5.7	3.76
Waste Combustion	6.6	3.84
Other Combustion	5.9	5.07
Sewage Gas	5.9	-
Average	7.2	4.36

Source: Mitchell, 1996, p. 175.

Significantly, in response to requests by the RECs, two new clauses were included in the terms of the Third Order. The first clause is referred to as a 'levy out' clause, which stated that, should the Fossil fuel Levy be ceased during the life of the contract the RECs would not be liable to pay the difference between the premium price for renewables and the pool price. The second so-called 'supply out' clause stated that were renewable energy generation to exceed 25% of the RECs own supply business, the RECs would not have to take the renewable electricity. Although together these two clauses squarely place the burden of the risk renewable technology with the developers, they did help in overcoming problems associated with previous Orders and emphasised the potential of renewable electricity supply as promoted by NFFO-FFL (Mitchell, 1996, p. 176).

Table 4 Summary of the 1994 Renewables Order (NFFO-3) and Associated Contracts

Technology Band	No. of Schemes Contracted	No. of Schemes Commissions as at 30/9/97	Contracted Capacity(MW dnc)	Capacity Commissioned at 30/9/97 (MW dnc)	Lowest Price Contracted (p/kWh)	Weighted Average Price(p/kWh)	Highest Price Contracted (P/kWh)
Wind > 1.6MW	31	5	145.9	19.1	3.9	4.3	4.8
Wind < 1.6MW	24	5	19.7	4.1	4.4	5.2	5.9
Hydro	15	5	14.4	7.9	4.2	4.4	4.8
Landfill Gas	42	32	82.0	58.3	3.2	3.7	4.0
Municipal & General Industrial Waste	20	1	241.8	28.2	3.4	3.8	4.0
ECAFW – Gasification	3	0	19.0	0.0	8.4	8.6	8.7
ECAFW – Residual	6	0	103.8	0.0	4.9	5.0	5.2
Totals	141	48	626.9	117.6		4.3	

Source: Table 3, Annex G, *Renewable Energy Bulletin* No. 7, 25 November 1997.

The NFFO-4

The details of the NFFO fourth tranche were announced in November 1995 and the Order should cover the period from early 1997 to 2016 (*i.e.* nineteen years). It will secure up to 843MW of new capacity, and therefore represents at increase of 186MW over NFFO-3.

There were further changes in the structure of the technology bands between NFFO-3 and NFFO-4. Roughly replacing the bands for municipal and general industrial waste, new bands were included for waste powered

CHP and waste powered fluidised bed combustion (FBC). It is unclear to what extent the figures provide useful comparison, but the share of total contracted capacity accounted for by waste powered CHP and FBC together in NFFO-4 was 38.6% (20.6% and 13.7% respectively) was exactly the same as the share accounted for municipal and general industrial waste in the previous Order. Similarly the ECAFW bands of NFFO-3 were replaced by new bands for anaerobic digestion of agricultural wastes and biomass gasification or pyrolysis. Together these new bands accounted for 8.8% of total contracted capacity (0.8% and 8.0% respectively) as compared with a 19.6% share achieved by the two ECAFW bands in NFFO-3. The wind power bands were again revised to aid smaller generators, the bands now being divided into less than 0.768MW dnc and greater than 0.768 MW dnc. The share of all wind power in total contracted capacity increased significantly between NFFO-3 and NFFO-4, rising by 14% percentage points to 40.4%, representing an increase in absolute terms of just over 175MW dnc to 340MW dnc.

Beyond this, though, there were no changes in procedure and the results of the contracts under the Order are presented in Table 4.5 below.

Table 5   Summary of the 1997 Renewables Order (NFFO-4) and Associated Contracts

Technology Band	No. of Schemes Contracted	Contracted Capacity (MW dnc)	Lowest Price Contracted (p/kWh)	Weighted Average Price* (p/kWh)	Highest Price Contracted (P/kWh)
Landfill Gas	70	173.68	2.80	3.01	3.20
Waste-Fired CHP**	10	115.29	2.79	3.23	3.40
Waste-Fired FBC***	6	125.93	2.66	2.75	2.80
Hydro	31	13.22	3.80	4.25	4.40
Wind > 0.768 MW	48	330.36	3.11	3.53	3.80
Wind < 0.768 MW	17	10.33	4.09	4.57	4.95
Anaerobic Digestion of Agricultural Wastes	6	.58	5.10	5.17	5.20
Biomass Gasification or Pyrolysis	7	67.33	5.49	5.51	5.79
Totals	195	842.72	-	3.46	-

Source: Table 4. *Renewable Energy Bulletin* No. 7, 25 November 1997, Annex G.
* Average price to be paid, weighted according to the expected output from each project.
** Combined Heat and Power,
*** Fluidised Bed Combustion

The NFFO-5

On 25 November 1997, the current Minister for Science, Energy and Industry, John Battle, issued the announcement of the fifth tranche of the NFFO coming into effect in 1998. Given lead times of up to five years and premium contracts of 15 years duration, this Order will run until 2018. The contracts awarded under NFFO-5 were announced on 19 October 1998, and total 261 projects amounting to almost 120,000MW dnc.

Although broad results are available (see below), it should be noted that at the time of writing detailed information on the awards is not available.[10]

Table 6   Summary of the 1998 Renewables Order (NFFO-5) and Associated Contracts

Type of technology	Contracted capacity (MW)
Landfill	310.814
Municipal and industrial waste	415.748
Municipal and industrial waste with CHP	69.971
Small scale hydro	7.38
Large scale wind (0.995MW or over)	340.161
Small scale wind (less than 0.995 MW)	28.672
Total	1172.746

Source: DTI Press Release, 'Going for Green - John Battle announces a real boost for green energy', 19 October 1998.

There had been much speculation that the fifth competition they might include new bands for: lands fill gas from existing sites used as tips before 25 November 1997; municipal and industrial waste; small scale hydro from stations with total capacity less that 5MW dnc; on-shore wind energy, split into small clusters/single machines not more than 1MW dnc, and other windfarms. However, the instigation of these new bands depended on there being sufficient quality applications to justify them, and the only changes made to technology bands in this Order were the establishment of bands for municipal industrial waste and CHP (both redesignated from the previous 'waste-fired' band), and the increase of the threshold between large and small scale wind power from 0.786MW to 0.995MW.

Continuing problems

In spite of the evolution of the non-fossil fuel obligation detailed above, there remain certain problems with the system. The most persistent criticism of the system has been that, even after the increases, the funds provided are simply too small really to assist these pioneer technologies, and that as a result the system has enjoyed most success in ensuring the development of projects which had few hurdles to overcome in the first place, such as those involving landfill gas, municipal and industrial waste-fired power or sewage gas.

The increased rate of the fossil fuel levy resulted in £1,324m collected in 1991/92, of which £1,311 went to nuclear generators, and the remaining £13m - about 1% of the total - to the renewables. As one commentator[11] has said, 'it does seem clear that the (previous) UK Government's tie up of renewables and nuclear in the non-fossil fuel obligation and the fossil fuel levy, has proved very unfortunate for renewables'. Since the financial year 1993/94 the decline in the sums received by nuclear generators[12] meant that the amount of money paid to renewables increased from £24m in 1992/93 to £99m, or 8% of levy funds, in 1994/95 a much greater increase than that Professor Littlechild announced in the OFFER *Annual Report 1993*, which predicted an increase in renewable funding over this period from 3% of the total levy payments to 6%.[13] This unexpected increase in renewables funding could be explained by the increased concern over climate change, together with low fuel prices which meant that renewables could no longer be expected to develop without government support. However, as Table 4.6 (above), clearly shows the amounts received by renewables, in absolute terms and as a proportion of levy funds, remained rather small.

The process of fossil fuel levy fund allocation as currently employed is also subject to much criticism. The method of competitive bidding, while it ensures that prices are kept down, also ensures that projects to refurbish existing schemes generally take precedence over new developments. In relation to wind technology, the competitive bidding process, tends to encourage the use of high wind speed sites, which are more efficient for electricity production, but which also tend to be more visible, remote, and therefore subject to planning problems. Also, because all projects must pay off their capital borrowings by the time their NFFO contract expires, any delay in commissioning can seriously jeopardise the viability of the project. Delays in obtaining planning permission, for example, are quite common, and have certainly lead to the demise of some renewable projects for precisely this reason.

Table 7   The amounts collected through fossil fuel levy, 1990/1991 - 1994/95

	1990/91	1991/92	1992/93	1993/94	1994/95
Rate (per cent)	10.6	11	11	10	10
The value of leviable electricity (£m)	11084.9	12036.4	12234.5	12340	12050
Amount collected (£m)	1175	1324	1348	1234	1205
Income from the levy (as reported in NE's annual report) (£M)	1195	1265	1280	1230	1251
Money actually paid to nuclear generators (£m)	1175	1311	1322	1166	1106
Money paid to the renewables (£m)	0	13	26	68	99

Source: OFFER's and Nuclear Electric's *Annual Reports*, OFFER's Press Releases, own calculations.

A further continuing problem with the NFFO system has been in its interaction with the planning system in England and Wales. Particular problems have been created by the tranche structure of the NFFO-FFL system, which resulted in a great many applications for planning permission at the beginning of each tranche competition, overwhelming some planning authorities and creating a fear that certain areas would be destroyed by an influx of new generation plants, in spite of the fact that many of those plants applying for permission would not receive assistance from NFFO-FFL and would not go ahead. Wind power was faced with particular problems, owing to the rather obvious nature of wind turbines and the concentration of suitable wind power sites in certain areas of the country (Mitchell, 1995).

The fundamental problem at the heart of this issue concerns the ultimate responsibility that rests with planners for the success of a much wider government policy, and the lack of guidance on that wider policy which planners receive. In December 1991 a 'draft planning guidance note' was issued and the final guidance note was published in February 1993 with the aim of better informing planners of the aims and methods of the government's policies towards renewables and of helping them to take this into account in their planning decisions. However, the guidance note is seriously flawed, underestimating the scale of the renewables generating plans which planners are likely to encounter and giving potentially contradictory guidance, for example on wind power. Although the announcement of the NFFO-5 explicitly states that developers are required to '...consult widely on their project proposals so that their projects meet the concerns of the community likely to be expected,' (*Renewable Energy Bulletin* no. 7, 25 November 1997, p. 3) this is unlikely to do much to solve the planning problems. It may simply be that only a radical change in local opinion, particularly among the agricultural community, can bring about the kind of changes in planning consent that will be required for the UK's renewables potential to be fulfilled.

5. Conclusion

The current system of support for renewables generated in England and Wales, that based on the non-fossil fuel obligation and compensation for premium prices for 'renewable electricity' via the fossil fuel levy has certainly enjoyed a degree of success since its inception in 1990; there is almost certainly a greater amount of electricity generated from renewable sources than there otherwise would have been. However, the coupling of support for renewables with - far more sizeable - support for nuclear energy, together with all the problems experienced with the way in which the Orders and the Levy have been handled, have meant that renewables support fell a long way short of what it might, realistically, have been.

Now that the nuclear element of the fossil fuel levy has been discontinued, and the UK nuclear industry is showing operating profits there is a clear opportunity to review the functioning of renewables support in the light of this change in context. If one accepts that there is a case for government support for renewable generation on positive externality grounds then there are several options financing available to government, perhaps the most attractive of which, especially on grounds of simplicity, is that of general taxation. However, even if one sees the role of government rather as 'enabler' than 'provider' then there does appear to be a feasible market solution that could be attempted.

At the time of writing a government bill on the non-fossil fuel obligation is progressing through Parliament, and if passed (and it seems that it will be) it will provide the basis for a continuance of the current NFFO-FFL for renewables financing. In this instance maintenance of the status quo is undoubtedly the easy option, and it is not difficult to see why a government with such an ambitious list of projects for its time in office would seek to minimise complications in other areas. However, there is a hope that in the context of the Government's current review of energy policy as a whole there will be serious consideration of the position of renewables in electricity generation and of methods for their assistance, and perhaps will produce a radical and effective new policy in this area.

References

Boulter, H. (1996), *Inside Sellafield*, Pinter, London.
Corry, D., Hewitt, C. and Tindale, S. (1996), *Energy '98, Competing for Power*, IPPR, London.
Department of the Environment and Welsh Office (1993), Planning Policy Guidance Note 22, February 1993.
DTI, 'Renewable Energy Bulletin' 1-7, DTI, London.
DTI (1994), 'Wardle Makes Third Renewable Energy Order', Press Release DTI, 20 December.
Grenville, A. (1992), *Journal of Energy and Natural Resources Law*, vol. 10, no. 2, pp. 285-292.
House of Commons (1992), *Renewable Energy*, vol. 2.
International Institute for Energy Conservation (1997) *The Potential for Sustainable Energy in the UK, a survey of recent literature for Greenpeace UK*', IIEC-Europe, London.
Jasinski, P. (1997), *The Nuclear Industry and the Fossil Fuel Levy in England and Wales*, Oxford Economic Consulting Ltd., Oxford.
Mitchell, C. (1995), *Renewable Energy in the UK: Financing Options for the Future*, Council for the Protection of Rural England, London.
Mitchell, C. (1996), 'Renewable Generation - Success Story?' in Surrey, J. *The British Electricity Experiment, Privatization: the record, the issues, the lessons*, Earthscan, London.
National Audit Office (1994), 'The Renewable Energy Research Development and Demonstration Programme', HMSO, London.
Roberts, J., Elliott, D. and Houghton, T. (1991), *Privatising Electricity. The Politics of Power*, Belhaven Press, London.
Surrey, J. (1996),*The British Electricity Experiment, Privatization: the record, the issues, the lessons*, Earthscan, London.

Notes

1. Public electricity suppliers are first tier electricity suppliers, and currently all first tier electricity suppliers are RECs. Thus in this paper 'RECs' is often used for PESs.
2. See, A. Grenville (1992) p.286. See also, Electricity Act 1989 Section 33 (8).
3. See, A. Grenville *op. cit.* p. 289.
4. See: John Hunt, Environment White Paper: Blueprint for environmental charges ostponed, *Financial Times*, 26 September 1990.
5. Roberts, Jane *et al.*, *op. cit.*, p. 151.
6. *Renewable Energy Bulletin*, No 6, December 1995, p. 1.
7. A fascinating account of such nuclear 'incidents' is found in Boulter (1996).
8. Roberts, Jane *et al*, *op. cit.* p. 151.
9. Quoted in Mitchell (1995) p. 17.
10. It is expected that the *Renewable Energy Bulletin* for November 1998 will contain more detail.
11. Roberts, J. (1991).
12. The data for 1995/96 were not yet available in the moment of writing this report.
13. In the previous year the renewables received 3% instead of the predicted 4% of the money collected.

5 Energy Efficiency in Germany

WOLFGANG PFAFFENBERGER, CHRISTOPH OTTE

1. Energy efficiency - a new old issue

After a break of several years, energy efficiency returns to be an issue of scientific and political discussion. In contrast to the seventies, the new meaning energy efficiency has gained recently, cannot be reduced to the impulse of energy prices. Crude oil prices are on the same level as ten years ago in nominal terms. In real terms, there has even been a price reduction.

On the other hand, the problem of exhausted fossil energy resources is not pressing at the moment because every year, new deposits are being explored with a volume that exceeds the annual energy consumption.

Another argument of the past, namely the increase of energy efficiency as an answer to the threat to western industrial countries, to be dependant on the OPEC states, especially on the oil exporting countries of the near and middle east, has become obsolete due to a successful diversification on the import side. In 1997 only about 30% of German crude oil imports came from OPEC-member states. The biggest share with roughly 40% were imports of British and Norwegian North Sea oil, another 25% came from Russia.[1]

One can say, the energy efficiency policy of the seventies and early eighties emerged from a simple supply-demand calculation: the consumers reacted to a shortage/rise in price of energy supply with measures that enabled them to reduce their energy demand. Today, the issue of energy efficiency must be seen in a totally different context. The discovery of the greenhouse-effect and the decisive role that CO_2-emissions from burning fossil energy sources play contribute to a 'rebirth' of energy efficiency policy measures as a part of the climate policy instruments. Here, great importance is attached to them because global energy supply will continue to be based on carbon containing energy sources and a CO_2 abatement technology has not yet been developed.

Recently, the context in which the issue of energy efficiency is being discussed, again has changed. Since the Conference of the United

Nations for Environment and Development held in 1992 in Rio de Janeiro the new generic term 'sustainable development' is developing towards a social model and a standard of political action.

In general, the term 'sustainable development' includes questions of the long-term survival and living together of mankind. It means searching for strategies, life styles and production processes that stabilize the relations between human beings and environment, between developed economies and developing countries and between present and future generations on a level that satisfies all parties. These developments have led to different demands and objectives energy efficiency policy has to meet nowadays than ten or twenty years ago:

- The main objective of an energy efficiency policy no longer only is safeguarding the energy supply of western industrial countries but also a sufficiently long-term supply of energy, taking into account ecological soundness, inter-temporary and geographical justice.

- An energy efficiency policy first of all has to be judged by its contribution to reduction of emissions, especially CO_2.

- An internationally focused energy efficiency policy must take care that the 'energy hunger' of industrial countries does not impair the chances for development of the rest of the world and that the increasing energy demand of developing countries is being matched in an environmentally acceptable way.

- Energy efficiency means that certain energy services are provided at a minimum energy use. In the course of the topical discussion about sustainability of life styles and patterns of consumption, it is also the question, if an increase of energy efficiency is sufficient or if it is necessary to conserve energy, that is to give up on certain energy services completely.

2. Energy efficiency potentials in Germany

2.1. Starting point: Consumption of primary energy in the status quo

The primary energy consumption in Germany in 1997 was 14,490 PJ. Table 1 shows the distribution on the different energy sources.

Table 1 Primary energy consumption in Germany 1997

Source of energy	PJ	Share (%)
Crude oil	5727	39.5
Natural Gas	2984	20.6
Hard coal	2043	14.1
Lignite	1591	11.0
Nuclear energy	1858	12.8
Hydro and Wind Power	73	0.5
Foreign Trade Balance Electricity	-9	0.0
Others	223	1.5
Σ	14,490	100.0

Source: Arbeitsgemeinschaft Energiebilanzen

Relating these consumption numbers to the population of 82.1 Million persons, a *per capita* consumption of 176 GJ results. The German *per capita* primary energy consumption ranges below average of G7- and OECD-member states. Germany has a share of 4.2% of the global primary energy consumption while its share of world population is only 1.4%.

Regarding the overall economic energy intensity, that is the relation of primary energy consumption to GDP in real terms, the picture alters. The German GDP was 3,126.7 billion DM in 1997 (in prices of 1991). For the above defined efficiency indicator this means a number of 4.63 GJ/1000 DM (1991) GDP. The development of the last years shows significantly an uncoupling of energy consumption and national economic development. GDP (in prices of 1991) grew by 9.5% from 1991 to 1997, while primary energy consumption remained more or less constant. So on this comparatively highly aggregated economic level a continuous increase of energy efficiency could be seen. However this increase has slowed down recently.

In spring 1998 the Federal Government has introduced an eco-political priority programme. For the field of energy use, the Federal Government's guideline demands: doubling of energy productivity until 2020 on the grounds of 1990.[2] In 1990 this number was 188 Million DM (1991)/PJ, so for 2020 an energy productivity of 377 Million DM (1991)/PJ has to be achieved.[3] Last year with one PJ of primary energy use a volume of 216 Million DM GDP (in prices of 1991) could be produced.

2.2. From primary to final energy: Efficiency on the first level of transformation

Table 2 gives an overview about the scales of primary and final energy consumption in Germany in the year of 1996.

Table 2 Primary and final energy consumption in Germany 1996

	PJ	Share (%)
Primary energy consumption	14768	100.0
Consumption and losses of the energy sector	4138	28.0
Non-energetic consumption	1002	6.8
Final energy consumption	9628	65.2

Source: Arbeitsgemeinschaft Energiebilanzen

Relating primary to final energy consumption results in the national energy economic efficiency for this stage of transformation, which was 65.2% in 1996. The number is more expressive when correcting primary energy consumption by the non-energetic consumption, so that the latter will not erroneously debit the efficiency of the transformation sector. This adjusted efficiency was 69.9% in 1996 and has increased by about 2 percentage points since 1990.

This paper focusing on the efficiency of final energy and not primary energy use, the different fields and technologies of transformation are not analysed in detail. The analysis of final energy use in the next chapter is more oriented by the sectors of consumption and by fields of application than by energy sources. As an exception from these principles an overview about production and use of electricity is inserted at this point. Table 3 shows production and use of electricity in Germany in 1997.

The gross electricity consumption in Germany was about 545 billion kWh in 1997, and thus ranged still below the number of 550 billion kWh, that had been measured in 1990, the year of the reunification. The breakdown of the East German economy, especially of industrial production, led to a decrease of electricity consumption in the range of 40% between 1989 and 1994.[4]

Table 3 Production and use of electricity in Germany 1997

	billion kWh
Domestic production by	
• Power plants of public utilities	485.0
• Industrial power plants	54.8
• Railway power plants	7.4
Gross domestic production	547.2
+ Electricity imports	37.5
- Electricity exports	39.9
Gross electricity consumption	544.8
- Internal consumption power plants	38.9
- Pumping electricity consumption	5.6
- Grid losses	20.4
Net electricity consumption	479.9

Sources: BMWi 1998, VDEW 1998

When comparing, as before with primary energy consumption, the trend of electricity consumption and GDP, a similar picture emerges: economic growth and the electricity consumption have been uncoupled. The overall economic electricity intensity decreased from 1991 to 1997 from 188.8 to 174.2 kWh/1.000 DM GDP (1991). This decrease with nearly 8% is not as strong as for primary energy intensity. The cause of this decline was, in addition to the growth of GDP, especially the increased energy efficiency in the industrial sector.

For analysing efficiency in the electricity producing sector itself, there are two main factors:

- Electricity consumption inside the sector, i.e. the internal consumption of power stations, pumping electricity consumption and the grid losses.
- The efficiency of transformation of different primary energy sources into electricity.

The first position, the internal consumption of the sector was about 12% of gross domestic production throughout the past years. Table 4 gives an overview about the efficiency of conventional heat power stations in Germany. The electricity production from nuclear energy, hydro and wind power is not listed, because based on the statistical systematic of the efficiency principle, an efficiency of 33% for nuclear energy and of 100% for hydro and wind can be assumed.

Table 4 Fuel consumption and gross electricity production 1996

Energy source	Fuel consumption for electricity generation (PJ)	Gross electricity generation (TWh)	Transformation efficiency (%)
Hard Coal	1397	152.8	39.3
Lignite	1426	144.3	36.4
Natural Gas	470	52.7	40.3
Heating Oil	65	6.8	37.6

Sources: BMWi 1998, own calculations[5]

Comparing these degrees of efficiency with the present standard of technology on the power plant market, a substantial technical efficiency potential is the result. According to manufacturers of gas fired combined cycle power plants, the electrical net efficiency is 58%. For hard coal power plants best efficiencies are currently at 45% and for lignite 43%.[6]

This technical potential can only be achieved with the construction of new power plants. Until the year of 2005 a fundamental renovation of capital stock owned by the electricity sector in West Germany cannot be expected because throughout the past years there have been considerable modernizations in the course of adjustment to tightened environmental demands. The power plant pool in the NFS has been fundamentally restructured and modernized since unification. In the past years several new lignite power plants and gasfired combined cycle power plants as well as one hard coal power plant have been constructed and on the other hand old GDR power plant capacities have been closed down. Thus the structure of electricity generation in East Germany will hardly change until 2020. Therefore increases in efficiency of power generation can only be expected from the year 2010 on, because then the power plant stock in the old federal states of Germany (OFS) will have to be replaced gradually. According to predictions these developments will in total lead to an average improvement of energy efficiency in the power plant sector of 0.5% p.a. for the period from 1997 to 2000.[7] All predictions that deal with the future power plant structure in Germany include a considerable element of uncertainty, namely the role of nuclear energy.

For the development of grid losses there are no estimates available. In East Germany increases of efficiency are still possible because parts of the grid have not yet been modernized. Also the medium and high voltage grids that usually have lower loss rates, so far have not been brought to the same standard as in the OFS. Liberalisation of electricity markets in Europe could in tendency lead to more electricity being transported through longer distances which again means an increase of grid losses.

2.3. From final energy to energy use: Efficiency on the second level of transformation

2.3.1. Structure of final energy consumption in Germany

Table 2 shows that final energy consumption in Germany in 1996 was 9,628 PJ. Thus it was in contrast to primary energy consumption again above the number of the year of unification (1990: 9,441 PJ). Certainly, final energy consumption has increased less than GDP since 1991, but more than primary energy consumption and gross electricity consumption. The latter points to the other final energy sources having higher rates of growth than electricity and that fields of application where electricity only plays a minor role, such as transport or room heating, gain importance for final energy demand. Table 5 contains the structure of final energy consumption by consumption sectors.

Without wanting to anticipate the analysis of the industrial sectors below, a few basic trends can be identified: The industrial final energy consumption decreases in absolute terms. The relative importance of energy consumption of the supply side of the economy, i.e. industry, trade and services, diminishes. The final energy consumption of private households and the traffic sector increases in absolute terms as well as in relative terms. These tendencies also are being supported by the results of analysis that goes further back in the past.[8]

Table 5 Sectoral distribution of final energy consumption in Germany

Sector	1996 Consumption (PJ)	Share (%)	1991 Consumption (PJ)	Share (%)	Change in Consumption 1991-1996 (%)
Industry	2400	24.9	2693	28.9	- 10.9
Trade, Services, Military	1694	17.6	1680	18.0	+ 0.8
Households	2934	30.5	2515	27.0	+ 16.7
Transport	2600	27.0	2430	26.1	+ 7.0
Σ	9628	100.0	9317	100.0	+ 3.3

Source: AG Energiebilanzen

Closing this chapter we will have a look at the final energy consumption of 2020. According to a study by prognos AG it will be as high as 1996 and will have a structure as follows in Table 6.

Table 6 Structure of final energy consumption in Germany in 2020

Energy Source	Share (%)	Sector	Share (%)
Hard Coal	2.7	Industry	28.9
Lignite	0.8	Trade, Services, Military	17.7
Crude Oil Products	45.1	Households	23.2
Gases	25.9	Transport Sector	30.2
Electricity	20.8		
Distance Heating	4.1		
Others	0.6		

Source: prognos 1996

On the whole, it can be expected that in coming decades the so called rebound-effect will continue to be observable: The continuing increases of energy efficiency in all fields of consumption will in total be 'eaten up' by the just as continuing increases of production and consumption.

2.3.2 The final energy consumption of industry

Industry has a share of 25% that is 2.400 PJ of German final energy consumption. But it contributes to gross value added with 33%. The development in the past few years can be divided into two stages. The noticeable decrease of GVA and energy consumption up to 1993 can be reduced to the massive closing down of energy intensive production plants in the NFS after reunification. At the end of the adjustment process in 1994 the long-term trend again started to set in. The final energy intensity of the industrial sector in 1996 was 3,085 MJ/1,000 DM GVA (in prices of 1991), that is 40% below the number of 1973.[9] This decrease can mainly be reduced to three factors:

1. *Reduction of specific energy use per produced unit* In industry a continuous modernisation of production plants takes place. The implementation of new production technologies often leads to savings of energy, even if these were not the true motives of the firms. Additionally, there were and there are investments that consciously aim at reduction of internal energy consumption.

2. *Inter-sectorial change of structure* A considerable shifting between different branches of industry can be observed. The group of relatively energy intensive basic industry and the industry of producer goods tend to loose importance. On the other hand the less energy intensive investment and consumer good industry gains importance.

3. *Intra-sectorial change of structure* The change of production range inside the different branches also has energy consumption reducing effects. This can be illustrated with the example of the chemical industry. A shift away from the energy intensive primary matter chemistry, e.g. the production of chlorine or ammonium, towards the less energy intensive production of high quality products such as pharmaceuticals, cosmetics and agricultural products, can be observed within the production range.

It is estimated, that the constant renovation of capital stock and the intra-sectorial change of structure contributed 70% and the inter-sectorial structural change 30% to the improvement of energy efficiency during the past 25 years.[10]

Inter- and intra-sectorial differences also are responsible for the industrial energy intensity in the NFS being three times as high as in the old.[11] The share of the whole manufacturing industry that consists of basic industry is higher in the east than in the west. Additionally, in the east, these industrial sectors have higher growth rates as the investment goods industry, so this kind of regional inter-sectorial structure currently is being strengthened. But also while comparing companies of the same sector, it can often been seen, that the product range in the NFS is more energy intensive than in the OFS.

The next point to be discussed is, for which applications in industry final energy is used, and how efficient the transformation from final energy to energy use is organized. For this, the following fields of application will be distinguished: Space heating, hot water, other process heat, mechanical energy, lighting.

The following Table shows the importance of the different categories in the industrial sector of the OFS and in West Germany in total:

Table 7   Structure of final energy consumption 1993

Field of Application	Industry OFS (%)	All sectors OFS (%)
Space Heating	10.4	32.3
Hot Water	0.7	5.2
Other Process Heat	67.4	21.5
Mechanical Energy	19.8	39.1
Lighting	1.6	1.9

Source: VDEW 1995

The Table shows that 2/3 of final energy consumption in industry is for the production of process heat. Thus this field of application carries considerably higher weight than in national average and offers the most promising starting point for efforts to increase efficiency in this sector. Statements about the efficiency of final energy use, i.e. the relation between final energy use and the resulting useful energy supply are being complicated by the poor quality of data. The Federal Government published the following data for the OFS for 1990.

Table 8   Efficiency of final energy use in West Germany 1990

Field of Application	Efficiency (supply of useful energy in % of final energy consumption)	
	Industry	all sectors
Process Heat	58.0	53.5
Space Heating	70.0	72.5
Mechanical Energy and Light	46.0	25.0
In total	56.6	47.3

Sources: AG Energiebilanzen, RWE-Anwendungstechnik

On the whole, final energy efficiency in the industrial sector is higher than in national average. As will be shown later, this is mostly because of the extreme low efficiency of the transport sector in supplying of mechanical energy. Nevertheless, 40% of the final energy needed by the industrial sector remains unused and will be lost as waste heat. As a general rule, the relation of primary energy to final energy to energy use is about 3:2:1 for an economy on the whole.

The industrial final energy consumption concentrates on few industries. 84% of consumption falls to the ten largest consumers. These, however, produce 60% of industrial gross value added. Especially remarkable is the role of the chemical and the iron and steel industry. Despite only contributing 10% to the economic performance, more than half the energy consumption falls to them.

On the whole, ESSO predicts increases in energy efficiency for the overall industrial sector in the OFS of 2% p.a. for the period from 1996 to 2010 respectively 1.7% p.a. from 1997 to 2020.

The study by prognos AG already mentioned in chapter 2.3.1, quantifies the final energy consumption of industry in the year 2020 with 2,777.5 PJ. This would mean an increase of about 15% compared to 1996. The grounds for this prediction is an estimated annual increase of industrial net production by 2.3% p.a. from 1992 to 2020. The specific energy consumption is expected to decline during this period by 43%, that is 1.3% p.a.

2.3.3 The final energy consumption in the transport sector

The final energy consumption of the transport sector in Germany was 2,600 PJ in 1996. This is equivalent to a share of 27% of total consumption. In 1973 these numbers were only 1,579 and 16.6%. This increase in consumption by 64% during the past 23 years is the strongest of all areas of consumption (industry – 37%, households + 25%, small consumers – 2%).

The transport sector is very heterogeneous. Transport of passengers and freight is provided by a large number of transport systems, which have different energy consumption characteristics and whose shares of the whole transport volume develop differently. In the following the modal split in passenger and freight transport is described.

Passenger transport

It must be differentiated between public passenger transport and transport by private car. The former is a combination of many means of transportation, e.g. railroad, tramway, subway, buses, aeroplane. The private transport is only car and motorbike traffic (including rental cars and taxis). Table 9 gives an overview on the means of transport and the transport performances in the course of 1997.

Table 9 Modal split of passenger transport in Germany 1997

Means of transport	Transported passengers in millions	Transport performance in billions of passenger-kilometres
Public local passenger transport	9293.1	85.1
- railway transport	1585.2	33.5
- road transport	7707.9	51.6
Public long-distance passenger transport	323.1	83.1
- railway transport	145.4	31.0
- road transport	78.3	24.1
- air transport	99.4	28.0
Total public passenger transport	9616.1	168.2
Private passenger transport	49974.0	747.8
Total passenger transport	59590.1	916.0

Source: ifo 1998

Transport by private car plays a decisive role in passenger transport. Its share of passenger-kilometres is more than 81%. The public passenger transport splits into 38% for railroad, 45% for road traffic and 17% for air transport. Structure and volume of passenger transport have changed considerably during the past years. In general, there was a great tendency towards private transport and air transport to be seen in the modal split. Public local passenger transport and railroad transport lost - even in absolute terms - importance. Especially significant is the development in the NFS. In the former GDR, public passenger transport (excluding air transport) in 1987 had a share of over 40% of the total passenger transport. In 1992 this number fell under 19%, while the total passenger transport remained constant.[12]

Freight transport

Freight transport can be divided into local area and long-distance transport. While local transport is exclusively done by truck, for long-distance there are many means of transportation. Table 10 contains an overview about the freight transport in 1997 in Germany.

Table 10 Modal split of freight transport in Germany 1997

Means of transport	Transported freight volume in million tonnes	Transport performance in billion tonne-kilometres
Railway transport	316.8	72.8
Inland navigation	233.2	63.0
Road transport	872.3	232.5
Air freight transport	2.0	0.5
Pipelines	87.4	13.3
Total long-distance freight transport	1511.7	382.2
Road local freight transport	2346.7	67.8
Total freight transport	3858.4	450.0

Source: ifo 1998

Analogue to passenger transport also most of the freight transport is being carried out by road transport. Local and long-distance road freight traffic are about 2/3 of total freight transport. 16% respectively 14% of transport performance are carried out by railroad and inland navigation. Volume and handling of freight transport have changed considerably recently, again, there are parallels to passenger transport. In 1980 the total volume of traffic was 339.6 billion tonne-kilometres, road transport contributing a share of about 43%. The share of inland navigation being only slightly higher than today, railroad transport still had a share of over 35%. For long-distance freight transport there has been a strong shift from rail to road. In 1997 railroad only had a share of 60% of the freight transport volume of 1980, while road transport grew by 160%.

In the next step, it is to be analysed, how much energy the different means of transport need for supplying their volume of transport. Generally, the efficiency of final energy use in the transport sector is very low. Merely 18% of consumed final energy is available as useful energy. In national average, this rate is about 47%.[13] The above described modal split naturally plays a decisive role for this low energy efficiency, and especially the dominance of road transport and the poor exploitation of energy by combustion engines. In the following Table 11 the final energy consumption therefore is further splitted.

Table 11 Final energy consumption of the transport sector 1993

Area of transport	Final energy consumption (PJ)	Share (%)
Road transport		
- Private passenger transport	1588	61.15
- Public passenger transport	43	1.66
- Freight transport	628	24.18
Railway transport	89	3.43
Air transport	218	8.39
Inland Navigation	31	1.19
Σ	2597	100.00

Sources: Schiffer 1997, own calculations

With a share of 87% of the total final energy consumption of the transport sector, road transport has a dominant position. Motorized private transport contributes already 61% to the total consumption. Thus it consumes 37 times as much as the public passenger road transport despite only making up for the 10-fold passenger-kilometres.[14] The specific energy consumption of private passenger transport is about 2.12 PJ/billion passenger-kilometres, of public passenger transport it is 0.57 PJ/million passenger-kilometres. It is noticeable that air transport only makes up for 3% of total passenger transport and only for 0.1% of freight transport but consumes 8% of the total energy in the transport sector. Conversely, the share of railroad transport is 7% of passenger transport and 16% of freight transport but of energy consumption only 3%. There can be no doubt, that the main energy consumer in freight transport is road transport.

As can be seen from Table 11, private passenger transport is very important for energy consumption, therefore it should be taken a closer look at. Private passenger transport contains nearly exclusively transport by car. Table 12 contains some central keys of car transport in Germany.

Table 12   Data about car transport in Germany 1997

Key	
Car stock	41.5 million cars
- petrol	- 35.7 million cars
- diesel	- 5.8 million cars
Annual kilometric performance	
- Petrol car	12,000 km / year
- Diesel car	16,700 km / year
Specific Fuel consumption	
- Petrol car	8.9 l / 100 km
- Diesel car	7.4 l / 100 km
Total driving performance	525.26 billion car-kilometres

Sources: ESSO 1997, ifo 1998

The car stock has risen considerably during the last years. In 1970 there were only 15.1 million cars in Germany, in 1980 it was 25.9 million and 35.3 million in 1990.[15] But not only the number also the composition of car stock has changed. The trend went towards bigger, heavier and better performing vehicles. In 1973 still 50% of new cars registered had a cubic capacity below 1500 cm^3, 1995 only about 30%.[16] This trend only led to gradual improvements in energy efficiency of car stock. The average consumption (petrol and diesel) decreased in the period of 1973 to 1997 from 10.7 l/100 km to 8.8 l/100 km. That is only 18% in 24 years time. The ESSO AG quantifies the improvements of energy efficiency for the period from 1971 to 1996 from petrol cars at 0.7% p.a. and for diesel cars at 1.1% p.a.[17]

How are things looking for energy efficiency of different means of freight transport? According to an inquiry of the University of Heidelberg[18] the specific energy consumption of road freight transport was 2.48 PJ/billion tonne-kilometres in 1990. So, compared to 1960 energy efficiency has risen by only 13%. The development of energy efficiency was positively influenced by the drop of work transportation and increase of long-distance freight traffic as well as technical improvement. In contrast, the railway freight transport has a considerably lower specific energy consumption, namely 0.52 PJ/billion tonne-kilometres. From 1960 to 1990 energy efficiency grew by 73%. The reasons are the utilisation of modern locomotives, a larger average transport distance and the reduction of shunting. The lowest specific energy consumption is with inland navigation with 0.47 PJ/billion tonne-kilometres. Here, energy efficiency has increased by 50% since 1960. Reasons are the modernisation of the fleet, the extension of waterways and a risen capacity utilisation. Fact is, that with freight transport the specific energy consumption of transport by rail and inland navigation is only about 1/5 of road transport energy consumption.

Table 13 contains the forecast of prognos AG for the final energy use of the transport sector in 2020.

Table 13 Final energy consumption of transport in 2020

Area of transport	Final energy consumption (PJ)	Share (%)	Change compared to 1993 (Table 11) in %
Road transport	2433	83.64	+ 7.7
Railroad transport	102	3.51	+ 14.6
Air transport	313	10.76	+ 43.6
Inland navigation	61	2.10	+ 96.8
Σ	2909	100.00	+ 12.0

Sources: prognos 1996, own calculations

For their results prognos used the following partial forecasts:

- Car stock 2020: 50 million vehicles;

- Average annual kilometric performance 2020: petrol car: 12,500 km, diesel car: 15,000 km;

- Specific fuel consumption in 2020: petrol car: 6.5 l/100 km, diesel car: 5.6 l/100 km;

- Doubling of transport performance in road freight and air transport from 1994 to 2020;

- Increase of carriage performance from 1994 to 2020 in rail passenger transport by 25% and in rail freight transport by 100%.

2.3.4 Final energy consumption in the household sector

The final energy consumption of the household sector was 2,934 PJ in 1996. It had developed very dynamically since reunification and was in 1996 23% higher than in 1990. With a share of over 30% the household sector is the biggest field of consumption in Germany. It was not until 1993 that the sector gained this position. The household sector contains mainly the energy use that is related to housing. This includes the energy need for the supply of space heating and process heating (cooking and baking), hot water, the supply of mechanical energy (household appliances, information-, home entertainment- and telecommunication technology) and for lighting. Volume and structure of final energy consumption of the household sector are mainly determined by the following four factors: Demography. housing situation, equipment with household appliances, behaviour.

At the end of 1996 82 million persons lived in 37.281 million households in Germany. This means that the average household consists of 2.2 persons. There have been considerable structural shifts in the private household sector during the past years and decades (see Table 14).

In 1996 2/3 of the German households consisted of one or two persons. The single person household is the most common form of households in Germany. In general, there is a trend towards smaller households. This development led to an over-proportional growth of households compared to population. These trends also increase the energy consumption of the household sector, because some indices, e.g. the number of flats, rather depend on the number of households than on the number of population.

Table 14 Trend of structure of private households in Germany

Year	Population (Million)	Households (Million)	Share of households with ... persons (%)				Persons per household
			1	2	3 and 4	5 and more	
1950	49.850	16.650	19.4	25.3	39.2	16.1	2.99
1970	60.176	21.991	25.1	27.1	34.9	12.9	2.74
1980	61.481	24.811	30.2	28.7	32.3	8.8	2.48
1991	80.152	35.256	33.6	30.8	30.5	5.1	2.27
1996	82.069	37.281	35.4	32.3	27.7	4.6	2.20

Source: Bundesbauministerium 1998

How do Germans live ? At the end of 1995 there were 35.954 million flats in Germany. The following Table 15 contains some specifications of housing supply in West and East Germany.

Table 15   Housing supply in West and East Germany 1995

Keys	OFS	NFS	Germany
Floor area per flat (m^2)	87.0	69.7	83.6
Floor area per inhabitant (m^2)	37.9	31.8	36.7
Rooms per flat	4.4	4.0	4.4

Source: Bundesbauministerium 1998

These keys are of great importance for energy consumption. So, the dimensions of a flat is important for the demand for space heating and the number of rooms for the consumption for lighting purposes.

Another relevant specification for energy consumption is the age-distribution of the housing stock, Table 16 contains an overview.

Table 16   Structure of age of housing stock 1993

Year of construction	Number of flats (million)	Share (%)
Up to 1900	3.662	10.9
1901 - 1918	2.857	8.5
1919 - 1948	4.865	14.4
1949 - 1968	10.770	31.9
1969 - 1978	6.057	18.0
1979 - 1987	3.943	11.7
Since 1988	1.583	4.7
∑	33.737	100.0

Source: Bundesbauministerium 1998

As can bee seen, about 2/3 of all flats have been built after 1948. However, here are big differences between West and East Germany. While in the OFS about 70% of all flats have been erected after 1948, it is only 50% in the new ones. So the housing stock is comparatively young in West Germany, but only less than 20% of all flats have been built after the introduction of the first heat insulation regulation in 1978.

As a last thing, the heating structure has to be analysed. It can be differentiated by the types of heating system and the used fuels.

Also for the type of heating there are considerable differences between West and East Germany. In the OFS, and thus in whole Germany, central heating systems dominate, while in the NFS only one fourth of all flats is equipped with these systems. Here, single and multiple room stoves and district heating are on top of the list. In the long-term range, central heating systems will probably gain importance in whole Germany. It is hard to predict which effect this will have on energy consumption, because on one hand central heating systems have a higher technical efficiency than single or multiple room stoves. On the other hand, the higher operating convenience could influence the utilisation behaviour towards higher heating energy consumption.

Most central heating systems in Germany still are fired with heating oil, which in the future is expected to be gradually replaced by gas. This can already be observed in the NFS, where the newly installed central heating systems are mainly fired with gas. For single storey heating systems gas dominates clearly. This will deepen throughout the next years, when the coal fired single storey heating systems in East Germany will be changed. On the whole, the heating structure is as follows:

Table 17 Heating structure of housing stock and of new residential buildings 1996

Energy source	Germany (%)	OFS (%)	NFS (%)	Heating in new residential buildings in Germany 1996 (%)
Gas	39	41	33	72
Heating oil	34	39	11	17
Electricity	6	7	3	1
Coal	9	5	25	-
District heating	12	8	28	10

Source: Schiffer 1997

This heating structure provides relatively favourable conditions for an energy efficiently organisation of the market for space heating. With central heating systems the highest degrees of efficiency can be realised with gas fired condensing value boilers, so growing gas shares in the future will have an increasing effect on energy efficiency. Oil heating systems based on condensing value technology also reach high degrees of efficiency. The disadvantage of the use of coal in space heating from the point of view of energy efficiency, will be reduced by exchanging coal fired heating plants and stoves especially in East Germany. The efficiency of supply of space heating by district heating systems will also increase because in this sector an increasing number of co-generation plants are used instead of mere heating plants.

In addition to demography and housing situation the equipment with household and other appliances is another factor that influences the final energy consumption of the household sector. This applies especially to the demand for electricity.

For some appliances such as refrigerator, washing machine and television set a high degree of saturation near 100% is already reached. A large part of newly bought appliances replace old ones and contribute to a decrease of final energy demand because of a lower specific electricity consumption. However, research has shown that old appliances often are not being disposed off, but continue to be used as a second or third appliance in the same household.[19] Thus the modernisation of the stock of electrical appliances paradoxically contributes to an increase of energy consumption. A potential of growth for the equipment of households remains, especially in the NFS, for dishwashers, microwave ovens and tumble dryers, that in average are missing in every second household. The highest percent increases are expected for communication, information and safety technology.

The last of the factors mentioned at the beginning of the chapter is the behaviour of residents. Here, energy consumption increasing as well as decreasing tendencies exist side by side, so it is hard to tell the overall direction of development. The VDEW forecasts that from the behaviour of households a slight increase of energy consumption can be expected.[20]

After looking at the influencing factors the energy consumption itself is to be examined (see Table 18).

The final energy consumption of private households is definitely dominated by space heating, more than 3/4 of the needed final energy in households goes to this field of application. Relating the energy consumption of 1,665 PJ, as mentioned in the table, to the total floor area of all flats in Germany the result is an average consumption of 170 kWh/m^2*a. This corresponds to a consumption of 17 m^3 gas or 17 l heating oil per m^2 and year.

Table 18 Structure of final energy consumption in the household sector 1993

Field of application	Consumption (PJ)	Share (%)
Space heating	1665	76.6
Hot water	261	12.0
Other process heat	85	3.9
Mechanical energy	129	5.9
Lighting	35	1.6
∑	2175	100.00

Source: Schiffer 1997

In the space heating sector there have been considerable increases in efficiency in the last few years. The specific oil consumption in l/m² was in 1995 in temperature corrected terms 40% below the level of 1970 for detached and semidetached houses in West Germany.[21] New buildings that correspond to the low energy house standard, however, only need 30-80 kWh/m² and year. The passive house, being a future development of the low energy house, needs no more than 15 kWh/m² per year. Because of the importance of the building stock, a main task for energy efficiency policy has to be reducing the energy consumption of old buildings.

Also for electrical appliances there have been relatively high increases in efficiency in the past. However, according to the Federal Office for Environment the losses arising from leaving electrical appliances in the stand by mode only, make up for 11% of the total electricity consumption of households.[22] Regarding the relation between final energy and energy use, an final energy efficiency of 65.2% for the household sector can be observed. This is the highest number of all consumption sectors and exceeds the national average by 18 percentage points. This relatively good result can mainly be attributed to the efficiency of space heating supply with 73%. The corresponding numbers are for the supply of process heat roughly 46% and for mechanical energy and lighting about 36%.

The prognos AG[23] expects a final energy consumption of 2,229 PJ for 2020, that is a drop by 24% compared to 1996. The different fields of application will have the following shares of consumption: space heating 77%, hot water 11%, process heat 2%, mechanical energy and lighting 10%. Specifically the following developments are forecasted:

- Increase of housing stock to 42.6 million in 2020;
- Rise of the average housing area to 94 m² in 2020;

- Improvements in heat insulation and in heating systems efficiency between 1992 and 2020 lead to a decrease of the surface specific heating consumption by 38%;
- The share of flats with central heating will be 99% in 2020;
- The heating structure in 2020 will be: gas 48%, heating oil 33%, district heating 10%, electricity 8% and coal 1%.

3. Energy efficiency policy in Germany

3.1. Legitimizing state energy efficiency policy

Government measures with the objective of energy saving have in Germany, like in other industrial countries, at first been a reaction to the oil price crises at the beginning of the 1970's. The energy saving act e.g. has been introduced in 1976. At that time governmental energy efficiency policy was being justified with the limited reach of fossil energy supplies and the fear of being dependant on the oil exporting states of Near and Middle East. Today, energy efficiency policy pursues other objectives. The resource problem not being as dramatic as it was considered at that times, energy saving and rational energy use today is seen more under the climate protection aspect as a possible promising CO_2-reduction strategy.

Situations of energy use are characterised by a certain relation between the factors energy and capital. The optimal input relation between energy and capital often cannot be realised.[24] A sub-optimal, because much to high, amount of energy continues to be used, although a partial substitution by capital is technically possible and economically reasonable. Different barriers that impede an efficient energy input are held responsible for this phenomenon, the so called 'efficiency gap',[25] therefore it is the task of energy efficiency policy to reduce these barriers.

German energy efficiency policy straight from the beginning was market focused, the decisive impulses for influencing consumer behaviour should come from energy prices. After the oil price drop in 1985/86 energy prices could no longer fulfil the stimulating role that politics attributed to them. The profitability of energy efficiency investments deteriorated. This led to governmental measures gaining importance regarding the energy efficiency objectives.

3.2. Instruments of national energy efficiency policy

Here, measures of German energy policy are described that directly aim at energy savings.

3.2.1. Information and advisory service for consumers

This generic term contains a large number of single measures, of which some examples are given below:

- Advisory service about saving and rational energy use in 330 towns.
- On-site advisory service for saving and rational energy use in housings. Housing owners can make use of the advice of engineers about heat insulation and heating system technologies.
- Support of energy saving advisory services in small and medium size enterprises. Industrial-, craft- and agricultural companies can get subsidies of up to 40% of the costs.

In the field of household appliances the Energy Labelling Act has been introduced in 1997. It is building the grounds for the implementation of two European Commission directives about the specification of the consumption of energy and other resources by household appliances and the demand regarding energy efficiency of household cooling and deep freezing appliances. In the meantime, the Federal Government has introduced two decrees: By the decree of energy labelling refrigerators, freezers, washing machines, tumble dryers and dishwashers have to be labelled with their electricity consumption and other environmentally relevant data. Likewise a decree of energy consumption maximum standards has been introduced, which lays down the maximum standards for new refrigerators and freezers for household use.

3.2.2. Financial subsidies and tax relieves

National financial aids at present concentrates on the NFS, hoping to obtain a higher energy saving effect with a certain input there than with subsidies used in the old part of Germany. These measures include for instance low interest loans granted within the programme of housing modernisation by the Credit Bank for Reconstruction (KfW). A great part of these loans (27%) had been used for financing energy saving measures, being carried out in relation with measures of maintenance which had been necessary anyway. On the whole, measures on 46% of the East German housing stock had been subsidised.

One main area during the last years was the district heating modernisation programme, with a volume of 1,2 billion DM. In the NFS 28% of all flats are being heated by district heating. With the aid of the rescue programme this relatively high share, which is only 8% in the OFS, was to be stabilised. This was made possible above all by supporting co-generation plants. Despite district heating being a main area of energy policy in the former GDR, co-generation only had an astonishing minor importance.[26]

For the OFS, the KfW started a supporting programme with a volume of 5 billion DM for CO_2-reduction in the housing stock in 1996. Until the end of 1997 measures at 161,000 flats had been financed by this programme. 37% of the money was used for heat insulation, 20% for the renovation of windows and 35% for the installation of condensing value boilers or low temperature boilers.

In the course of taxwise support of ownership of residential buildings, there is an additional support for energy saving investments ('eco-bonus'). It includes a subsidy of 500.00 DM p.a. for the maximum of 8 years for the installation of heat pumps, solar or heat recovery plants respectively of 400.00 DM p.a. for the construction of a low energy house.

3.2.3. Regulation by law

In addition to the above mentioned rather 'soft' instruments, there have also been some efficiency standards laid down by decree during the past 20 years. However, this was only done in two partial areas, namely utilizing of the by-product heat and in the building area.

3.2.3.1. Heat as a by-product

There are several technical processes, where the by-product heat is produced, which cannot directly be used and therefore will be radiated into the environment. The Federal Act on Immission Protection (BImSchG) obliges the operators of plants that are subject to authorisation, either to internally use the waste heat or to provide third parties with it, if technically possible and reasonable. This kind of utilization of heat has to be regulated by decree. As a concession to the self-obligation-statement made by the German industry with regard to climate protection, these measures have temporarily been put aside. Only for waste combustion plants a guideline exists, requiring the heat which is not usable internally or by third parties, to be used for electricity generation, if an electrical capacity of more than 500 kW is possible.

3.2.3.2. Housing area

Since the beginning of energy efficiency policy in the seventies the reduction of energy consumption in the sector of space heating is one of the main (regulated) areas. The legal basis is the already mentioned energy saving act. On that grounds, at present there are three decrees by which different approaches to energy saving are realized: Decree on Heat Insulation, Decree on Heating Systems, Decree on Heating Costs.

The fact that all demands laid down in these decrees have to be 'economically reasonable' illustrates, that the energy efficiency policy is basically market oriented. According to the Energy Saving Act this is the case when expenditures pay off by the energy savings resulting within the regular service life.

The decree on heat insulation aims at a limitation of annual heating demand for new buildings and for specific changes in construction at existing buildings. This happens by laying down standards about the reduction of transmission heat losses.

The decree on heat insulation concentrates on the need for heating of new buildings and thus only will lead to a slow improvement of energy efficiency in the whole housing stock. The average heating need for the total housing stock in Germany is 220 kWh/a per m^2 usable surface. Buildings that had been erected after the old decree on heat insulation of 1984, have a consumption of 80 to 190 kWh/ m^2*a. After its amendment in 1994 only a maximum annual heating demand between 55 and 100 kWh/m^2*a is allowed.

§ 8 of the decree on heat insulation contains regulations about the limitation of heating needs for existing buildings. At maintenance or modernisation of roofs and external walls and at renovation of windows, also measures for energy saving have to be applied. However, this rule is implemented only insufficiently. Here, the responsible federal states have to improve the execution.

The decree on heat insulation also serves to implementing of Art. 2 and 3 of the so called SAVE-guideline of the European Community. The introduction of a heating demand pass for new buildings laid down in the decree on heat insulation corresponds to the energy pass demanded in the guideline, which describes the energy related characteristics of a building and enables potential users to judge the energetic quality of the building and draw conclusions regarding the heating demand to be expected under standardized conditions. Here, it is taken into consideration to replace the heating demand pass by an energy pass, which is to be issued at the end of construction and does not contain planning numbers but shows the real energy consumption records. Additionally a similar certification could also be prescribed in the course of refurbishment of existing building.

Energy consumption not only depends on the applied heat insulation measures, but also on the efficiency of the heating system. In Germany heating systems have to match certain demands that are laid down in the decree on heating systems in order to limit these losses. The decree not only refers to new heating systems but also to the already used. Old systems that have been installed before the introduction of the first decree on heating systems in 1978 are usually dimensioned too large and have already reached the end of their service life. These machines had to be adjusted to the real heating needs until the end of 1997, graduated by rated capacity and age. Also, all central heating systems had to be equipped with modern control systems that allow the adjustment according to outside temperature and time until the end of 1997, and all radiators had to be equipped with thermostatic valves.

Because of the climatic situation in Germany, heating systems are exclusively running in partial load during heating seasons. The decree on heating systems therefore requires that only low temperature or condensing value boilers that have efficiency advantages compared to conventional boilers in this field of application are allowed to be activated since 1998.

For the future a reduction of annual heating demand for new buildings by 30% as against the currently valid decree on heat insulation is strived for. The relation between building and heating system gaining importance by this reduced heating demand, it is planned to combine the decrees on heat insulation and heating systems in a single so-called decree on energy savings. At present the exact structure of this decree is still being discussed but there are also critical voices doubting that the possible energy saving effects justify the costs related to the implementation of new standards.

By the European Community's SAVE-guideline of 1993, the EU-member states are obliged to take care that costs for heating, air conditioning and hot water are accounted according to consumption. In Germany this method of accounting has been introduced bindingly in 1981 by the decree on heating costs.

Since then the heating costs in buildings with several housing units had to be shared according to individually measured heating consumption in each flat. 50 to 70% of the operating costs of the heating systems, have to be accounted according to the measured consumption. The rest of the costs are allocated according to living space, usable surface or the enclosed space.

The objective of this decree is to achieve a reduction on energy consumption without large investments by changing the behaviour of the users of the building. By providing financial advantages, the individual monitoring of heating energy consumption gives an incentive to the users to save heating energy. Without monitoring, the supply of heating energy is comparable to a public good and the share of heating costs to a tax for financing this good. Such a construction naturally has the tendency to increase energy consumption because the single user is likely to consume as much as possible of the 'public good' for his share of 'tax payment'. It is estimated that the introduction of the decree on heating costs led to a decrease of energy consumption by 50%.

References

Blick durch die Wirtschaft (1997): 'Straßengüterverkehr trägt maßgeblich zum Treibhauseffekt bei', in *Blick durch die Wirtschaft*, 04.12.1997

BMWi (1998): Bundesministerium für Wirtschaft (ed.): *Energie Daten '97/'98*, Bonn 1998

BMWi (1998a): Bundesministerium für Wirtschaft: 'Rohölimporte Januar - Dezember 1997', in *BMWi-Tagesnachrichten* No. 10711, 16.02.98

Borch, Nickel (1998): Günter Borch, Michael Nickel: 'Marktforschung bei Haushaltskunden', in *Elektrizitätswirtschaft*, Vol. 97 (1998), No. 15, p. 31-35

BP (1996): BP Oil Deutschland GmbH (ed.): *Zahlen aus der Mineralölwirtschaft 1996*, Hamburg 1996

Bundesbauministerium (1998): Bundesministerium für Raumordnung, Bauwesen und Städtebau (ed.): *Haus und Wohnung im Spiegel der Statistik 1997/98*, Bonn 1998

Bundesumweltministerium (1994): Bundesministerium für Umwelt, Naturschutz und Reaktorsicherheit (ed.): *Erster Bericht der Regierung der Bundesrepublik Deutschland nach dem Rahmenübereinkommen der Vereinten Nationen über Klimaänderungen*, Bonn 1994

Bundesumweltministerium (1998): Bundesministerium für Umwelt, Naturschutz und Reaktorsicherheit (ed.): 'Entwurf eines umweltpolitischen Schwerpunktprogramms', in *Umwelt*, 1998, No. 5, p. I-XII

Ebersperger u.a. (1998): Ralf Ebersperger, Wolfgang Mauch, Cornelis Rasmussen, Ulrich Wagner: 'Wieviel Energie braucht ein Pkw?', in *Energiewirtschaftliche Tagesfragen*, Vol. 48 (1998), No. 5, p. 323-329

Esso (1995): ESSO AG (ed.): *Energieprognose 1995. Moderne Heizung - aktiver Klimaschutz*, Hamburg 1995

Esso (1996): ESSO AG (ed.): *Energieprognose 1996. Industrie verbraucht weniger Energie*, Hamburg 1996

Esso (1997): ESSO AG (ed.): *Energieprognose 1997. Mehr Strom aus Gas*, Hamburg 1997

FAZ (1998): 'Jedes zweite Altgerät bleibt im Haushalt', in *Frankfurter Allgemeine Zeitung*, 05.02.1998

Hensing, Pfaffenberger, Ströbele (1998): Ingo Hensing, Wolfgang Pfaffenberger, Wolfgang Ströbele: *Energiewirtschaft*, München 1998

ifo (1998): 'Verkehrsentwicklung 1998 im Zeichen des Konjunkturaufschwungs', in *Ifo Wirtschaftskonjunktur 2/98*

Laschke (1997): Bärbel Laschke: 'Dauerhaft höhere Energieintensität der ostdeutschen Industrie?', in *Energiewirtschaftliche Tagesfragen*, Vol. 47 (1997), No. 3. p. 147-151

OECD (1997): *OECD in Figures. Statistics on the Member Countries. Supplement to The OECD Observer*, No. 206, June/July 1997

Opitz, Pfaffenberger (1996): Petra Opitz, Wolfgang Pfaffenberger: *Verpaßte Stunde Null? Transformation am Beispiel der russischen Elektrizitätswirtschaft*, Münster 1996

Otte, Kuckshinrichs (1997): Christoph Otte, Wilhelm Kuckshinrichs: *Der globale Einsatz von Clean Coal Technologies zur Stromerzeugung: Sozioökonomische Effekte zusätzlicher Exportnachfrage für Deutschland*, Interner Bericht des Forschungszentrums Jülich FZJ-STE-IB-5/97, Jülich 1997

prognos (1996): prognos AG (ed.): *Die Energiemärkte Deutschlands im zusammenwachsenden Europa - Perspektiven bis zum Jahr 2020*, Stuttgart 1996

Schiffer (1997): Hans-Wilhelm Schiffer: *Energiemarkt Bundesrepublik Deutschland*, Köln 1997

Schiffer (1998): Hans-Wilhelm Schiffer: 'Deutscher Energiemarkt '97', in *Energiewirtschaftliche Tagesfragen*, Vol. 48 (1998), No. 3, p. 179-193

Statistisches Bundesamt (1997): Statistisches Bundesamt (ed.): *Volkswirtschaftliche Gesamtrechnungen*, Fachserie 18, Reihe 2, Input-Output-Tabellen 1993, Wiesbaden 1997

Statistisches Bundesamt (1997a): Statistisches Bundesamt (ed.): *Datenreport 1997*, Bonn 1997

Umweltbundesamt (1998): 'Vorstellung des Jahresbericht 1997 des Umweltbundesamt', in *Pressemitteilung*, No. 23/98

VDEW (1997): 'Haushalte nutzen Strom rationeller', in *Pressemeldung der Vereinigung Deutscher Elektrizitätswerke*, 16.07.1997

Weber (1997): Lukas Weber: 'Some reflections on barriers to the efficient use of energy', in *Energy Policy*, Vol. 25 (1997), No. 10, p. 833-835

Notes

1 See BMWI (1998a).
2 Energy productivity is expressed in GDP in relation to primary energy consumption. It is the reciprocal of energy intensity.
3 See Bundesumweltministerium (1998).
4 See Schiffer (1997).
5 1 PJ = 0.278 TWh.
6 For an overview see Otte, Kuckshinrichs (1997).
7 See ESSO (1997).
8 See Schiffer (1997).
9 See Schiffer (1997).
10 See ESSO (1996).
11 See Laschke (1997).

12	See Bundesumweltministerium (1994).
13	See Bundesumweltministerium (1994).
14	See Table 14.
15	See BP (1995).
16	See Schiffer (1997).
17	See ESSO (1997).
18	See Blick durch die Wirtschaft (1997).
19	According to a study by the Gesellschaft für Konsumforschung (GfK) 43% of all TV-sets and 64% of videos remain within the household (see FAZ from 5.2.1998).
20	See Borch, Nickel (1998).
21	See ESSO (1995).
22	See Umweltbundesamt (1998).
23	See prognos (1996).
24	See Hensing, Pfaffenberger, Ströbele (1998).
25	Weber identifies four classes of such 'barriers': institutional, market, organizational and behavioural barriers (see Weber 1997).
26	See Hensing, Pfaffenberger, Ströbele (1998).

6 Promotion of Renewable Energy in Germany

WOLFGANG SCHULZ

1. Introduction

The use of renewable energy (RE) in the past was the sole source of energy for a long time and only in the course of industrial development its role has diminished. In Germany the majority of the population is in favour of renewable energy supply. In the seventies this was caused by the wish for reduced dependency on imports and the consciousness about the limitation of the stock of fossil fuels. Later, environmental damage by the use of fossil fuels strengthened the point and in the meantime the danger of climatic change plays an important role. Also it is now better known that the biosphere reacts much more to the disregard of ecological principles than was believed before. In addition, that renewable energy is an alternative to nuclear energy and its potential dangers played also a role.

Taking all this into account, it seems surprising that renewable energy contributes only about 2% to the supply of primary energy. There has been no breakthrough, i.e. growing production of renewables has not altered the overall situation because of the general growth of the energy use.

Nevertheless there have been some successful developments from which can be understood which framework is positive for a fast diffusion of renewables. This paper will concentrate on such developments and analyze how barriers can be removed. In the beginning, the present contribution and the perspectives of renewable energy in Germany will be explained.

An increased role of renewables always implies that energy will also be used more efficiently. Rational use of energy has been treated in a different article.

2. Present use and potential of renewable energy (RE) in Germany

The basis of all renewable energy is solar radiation that can be used directly in solar collectors or photovoltaic cells or in the case of biomass, hydro power or wind, it is part of the natural cycles. Only geothermal energy is based on thermal processes in the inner part of the earth and tidal energy (not yet used in Germany) is based on gravitation of the moon.

Table 1 lists important technologies and their contribution. Figures are partly based on estimations because statistical coverage particularly regarding own consumption is not yet complete. Use of hydro power and fuel wood have been the most important forms of renewable energy in Germany for a long time. The growing tendency can be seen in the use of wind power, solar thermal energy and photovoltaic. The contribution of the direct use of solar energy is still on a very low level.

The seldom mentioned anaerobic processes in the treatment of waste water and sewage produce a much higher contribution than the use of solar energy, which is paid a lot of attention to. Also, the production of fuel oil on the basis of rape has reached a considerable level. In total in Germany at present about 23 GWh electricity, 35 PJ heat and 4 PJ fuel are produced from renewable sources. In some countries - like Denmark - the burning of waste is included into the statistic of renewable energy. As the calorific value of waste being usually from fossil origin here it will not be included in RE.

Table 1 Present use of renewable energy in Germany

Source	Transformation	Energy produced	Amount	Year	Tendency
Radiation	photovoltaic device	electricity	0.017 TWh/a	1996	growth 30% /a
Hydropower	turbine	electricity	19 TWh/a[1]	1996	constant
Wind power	wind energy converter	electricity	3.4 TWh/a[2]	1997	0.5 TWh/a increase
Biomass	thermal power station	electricity	0.15 TWh/a	1996	increasing
	or gasification + gas motor driven generator	heat	unknown		
Geothermal	geothermal collector	heat	0.5 PJ/a[3]	1996	unclear
Radiation	solar collector	heat	2 PJ/a	1996	20% /a growth
Biomass	heater	heat fuel 70 PJ/a	32 PJ/a	1996	
Environmental heat	heat pump	heat	low[4]	1997	constant
Biomass	anaerobic processes	fuel		1996	
	-biogas		0.2 PJ/a		slightly increasing
	-gas from sewage plants		5.5 PJ/a		
	-landfillgas		4.5 PJ/a		
		(produced electricity)	0.5 TWh/a)	1996	
	alcohol production	propellant	trivial	1991	constant
	rapeoil	propellant	3.4 PJ/a		constant

[1] about 4% of total electric production.
[2] 1766 MW*1943h/a (June 1997).
[3] also heat and power production possible
[4] about 50000 heat pumps in individial homes
Source: Langniß, 1997.

The technologies shown in Table 1 stand for a large number of single technologies. The use of biomass for instance includes:

- Burning of pieces of wood;
- Burning of wood chips;
- Production and burning of briquettes made of straw or wood;
- Various variants of burning straw bundles;

- Burning of rough sawdust for the production of heat or in large units for the production of heat and power (CHP) on the basis of steam turbines or motors;

- Or new attempts to use sterling motors for combined heat and power;

- Or many different concepts of gasification, pyrolysis or hydrolysis used as fuel for small motor driven generators;

- Or a plant for the production of synthetic gas to produce methanol which can serve as a motor fuel.

Solid biomass can either be a waste material of different origin or the product of plantations on the basis of plants delivering high output (willow, miscantus, grain). All these different forms cannot be explained in this article.

The existing perspective will therefore also be explained according to the rough division of Table 1. First of all we show the technical potential.

Various estimations of the technical potential have led to a great spectrum of results. We refer here to the results of the IKARUS- project which has completed a survey of the whole German energy sector in order to produce input materials for future scenarios.

The potentials of Table 2 can only partly be summed up because some technologies and utilizations of surface compete with each other. But taking into account, that electricity generation counts for 2.5 as primary energy and reduction of energy consumption being possible without renouncing to certain energy services, it seems theoretically possible to base the whole energy consumption in Germany on RE. However, there are disadvantages regarding economic aspects and security of supply connected to problems like handling and variation in supply of different sources of RE.

The relatively low price for fossil fuel is the most important obstacle for the development of renewable energy. There are only very few niches where renewables can compete with conventional energy sources.

Usually these are energy applications combined with other benefits (treatment of waste, which leads to the production of gas could) or applications which are meant for own consumption where the necessary labour input is not being valued, e.g. burning of wood pieces. Only hydro power can compete with conventional power at good locations. However most of the favourable locations are already in use. Often it is thought that wind converters of the 500 kW class are also competitive. There are however no long-term experiences with these converters yet that allow conclusions for the necessary maintenance costs for the plant life of twenty years.

Table 2 Technical potential of renewable energy in Germany

Source	Technical potential[1]	Energy produced
Photovoltaic		electricity
roofs 800 km²	140 TWh/a	
open spaces 3,500 km²	600 TWh/a	
Hydropower electricity		
< 1 MW	3 TWh/a	
> 1 MW	21 TWh/a	
Windpower		electricity
> 5 m/s	22 TWh/a	
< 5 m/s	49 TWh/a	
Geothermal	up to 600 PJ/a[2]	heat
Solar thermal		
425 km² of collectors	490 PJ/a	heat
Solid Biomass		fuel
energy plantations	30 PJ/a	
surplus straw/wood	210 PJ/a	
Anaerobic Processes		fuel
biogas	80 PJ/a	
sewage gas	28 PJ/a	
landfill gas	up to 22 PJ/a[3]	
Rapeoil	up to 18 PJ/a	propellant
Ethanol	up to 24 PJ/a	propellant

[1] present consumption of primary energy in Germany: 14,000 PJ/a (3900 TWh/a)
[2] would require 300 PJ of additional fuel
[3] declining after 2005
Source: Diekmann, 1995.

For some technologies the actual cost situation gives only little orientation for future mass applications. of scale and future development. One kWh of electricity produced by a photovoltaic cell in Germany costs about 2.0 DM and only large plants can reach a cost level of around 1.50 DM (electricity price for residential use at the moment is about 0.25 DM per kWh). Only because it can expected that through economies of scale costs could be reduced to something like 0.50 DM or less the potential of Table 2 is a realistic vision.

Table 3 Estimates about actual generation costs for RE, natural variation of supply and RE balances

	Present production costs	Availability fluctuations	Energy output/input
Electricity:			
Photovoltaic		day/night	4
roofs	1.90 DM/kWh$_e$	seasonal	
open spaces	~1.50 DM/kWh$_e$	dependent on weather	
Hydropower			120
available	>5 pf/kWh$_e$	only slight seasonal	
- new locations	~15 pf/kWh$_e$		
Wind energy			
(cost of electricity for private homes	25 pf/kWh$_e$)		37
Heat:			
geothermal	~22 DM/GJ	--	3
solar thermal	~100 DM/GJ	day/night, seasonal	14
solid biomass:			
energy plantations	>25 DM/GJ	stored energy	8-20
surplus wood/straw	>20 DM/GJ		20
(cost of electricity for private homes	30 DM/GJ)		
Fuel:			
Anaerobic processes		generally constant	
biogas	>20 DM/GJ		29
sewage gas	low[1]		
landfill gas	low[2]		
(natural gas costs per household	12 DM/GJ)		
Propellant:			
rape-oil	43 DM/GJ	stored energy	6
rape-metylester	50 DM/GJ		
ethanol	76 DM/GJ		1.3
(diesel oil costs incl. taxes:	33 DM/GJ)		

[1] distribution costs and treatment of sewage
[2] distribution costs, main reasons for degasification: safety and environmental protection
Source: Hartmann, 1995.

Table 3 gives a summary of the cost situation for renewables. Cost estimates differ widely and the Table does not give all the details Direct use of solar energy and use of wind power has the disadvantage that the supply is highly fluctuating. The uncoupling of consumption and production is possible by expensive storage systems. Alternatively, energy can be fed into a grid that can serve the same function. If the fluctuating contribution shoots over the limited regional capacity of the network, this could cause additional high costs for network improvement.

For Germany, concepts have been developed to use hydrogen instead of electricity as a renewable energy source. Hydrogen can be stored and transported without large losses. It then would be possible to produce it in areas with high solar radiation (like the Sahara) and to transport the energy through pipelines to Germany. This could also solve an additional disadvantage of renewables. The energy production per square meter is relatively low so that in a densely populated country like Germany a lot of conflicts about the use of the surface with groups representing interests like nature protection or commercial or residential use of the area is to be expected (Meissner, 1993).

Also, solid biomass is able to compensate for fluctuating supply of other renewables. At the moment such fuels are only economic if they can be used for base load purposes with 7,000 hours per year or more.

Table 2 shows that the technical potential for the production of compact biomass is relatively small. Allowing for imports, this contribution could be much higher if land use restrictions played only a minor role. For many arid and semiarid zones with low agricultural and forest use environmental benefits could be derived from such plantations which would also lead to structural improvements for these countries. Such a strategy which would also lead to a reduction of CO_2 through the plantations depends on an active strategy from the industrialized countries because those states do not have the economic power to create the necessary logistic structure.

For anaerobic digestion, the constant supply of resources and the costs for storage are an obstacle. For the use of sewage and landfill gas electrification typically can in most cases only be achieved without the use of heat because the location is usually far away from population.

As Table 3 shows, for some renewables the amount of indirect energy needed for the production of the plants is considerable. The production of ethanol, for instance, only leads to a net energy gain under certain circumstances. Also for geothermal energy, the heat pump using environmental energy (which is not included in Table 3) and photovoltaic energy production the input - output relation is not favourable.

3. Institutional conditions for renewable energy in Germany

Renewable energy needs improved institutional conditions if they are to achieve the importance required. To a large extent, this depends on the conditions created by political institutions, government and administration. We now analyze these conditions for individual renewable energies in Germany and evaluate their responsibility for the tendencies of development for these energies. The emphasis is on the present situation, we also look at the development and earlier obstacles which still can be felt today.

3.1. Past development

In this section we sketch out the developments of the past in order to find out the important factors that have promoted or impeded renewable energy. Therefore it is not necessary to state the trend for each renewable energy source.

In the seventies, policies towards renewable energies rather aimed at exporting to third world countries than implementing at home. At that time, partly due to the rising protest against nuclear energy, self-construction became relevant. These activities were however not promoted by official policies. Some self-constructing activities resulted in the development of professional companies that still have important shares in the market today. In certain areas (for example wood burning including logistics) offers from Scandinavia dominate the market. At the beginning of the eighties, some German companies producing agricultural appliances started to diversify into renewable energy. Experiences from self-constructing activities and from Scandinavia were integrated.

The pioneer work in the beginning was often hindered by negative attitudes of administration. Often also experts consulted by administrators made it difficult, partly because of lack of experience and partly because renewable energy in the dispute between nuclear and solar had an ideological touch. So it happened that highly developed wood burning appliances did not receive permission, were at the same time wood could be burned in furnaces meant for coal. As another extreme it was permitted that a large number of Danish furnaces for burning straw were built in which large stacks of straw could be burned. Because of the smoke from these stoves, burning of straw received a negative image.

But generally the market for renewable energy, particularly heat, developed quite well. This tendency ended however after 1985 with the fall of the oil and gas prices which led to a reduction by about 50% in fuel price. Well established firms that had emphasized renewable energy technology came into trouble or even went bankrupt. Technical developments ended. Now a phase of dependence on state promotion programmes began. These programmes were however not well geared to the needs of small and medium sized industry. The development only progressed with technologies that served the production of electricity. In the meantime highly developed motor combined heat and power stations became available in the market. In waste water and garbage processing increasing demands from resource and environment play a role.

Regarding wind power, in 1989 a change could be noticed. At this time in Denmark there was already a boom of wind energy converters. This is mainly attributable to a reliable framework regarding financial support and the pricing of the electricity produced.

In Germany in 1989 a far reaching programme began. At the beginning it was aimed at creating 100 MW wind power but was enlarged in 1991 to a capacity of 250 MW. In the first phase high investment subsidies were paid, where as in the second phase there was a choice between investment subsidy or subsidy of operation. In the beginning Danish manufacturers controlled the German market but slowly German companies began to develop so that in the meantime the largest share of converters is being produced in Germany (Becker, 1997). From 1988 to 1990 the installed capacity had already grown from 9.6 to 61.9 MW (Langniß, 1997). From 1991 a fixed price was set by a special law (Electricity Feed Law: Stromeinspeisegesetz), which increased the price by 8 to 10 pf/kWh. Under these conditions the 250 MW programme was already fulfilled in 1994 although at the end conditions had been narrowed in order to gain some experience with unfavourable locations away from the coastline. In the years 1993, 1994 and 1995 the capacity doubled from year to year so that in the meantime 2500 MW have been installed (Johnsen, 1998). For this boom not only the security of expectations but also changed structures are relevant:

- For financiers and insurance companies the risk becomes clearer with better experience;

- Capital funds are able to offer acceptable rates of return;

- Growing standardization simplifies administrative procedures for permission which also shortens planning time;

- The increased number of installation leads to reduction in maintenance costs (regional maintenance services);
- Investment cost is being reduced through increasing competition of manufacturers;
- Development activities increase so that it pays to have test institutions that contribute to durability and security of converters by certification;
- Increased experience by engineers leads to optimal design of converters;
- Consulting services improve so that most projects lead to success.

Since 1996 the awareness grew, that the amendment of the energy law, which became necessary to comply with a EU- framework guideline, would need an adjustment of the Electricity Feed Law. Intending a free competition, there was no space for area monopolies, so the Electricity Feed Law probably would loose its basis in the respect of the purchase duty of the particular utilities. Part of the utilities, who suffered most from high expenditures for wind electricity, considered this to be an opportunity to achieve a significant decrease of the legally fixed tariff paid for renewables or a repeal of the Electricity Feed Law. The corresponding political discussion led to a slow down in further construction activities and to a loss of confidence, respectively. The upward trend of annually newly built wind power plants stocked, and one big manufacturer went bankrupt.

Certainly, the Electricity Feed Law has survived the amendment of the energy law coming into effect in April 1998, but it was extended by some regulations, that contain a limitation of the share of electricity refunded in relation to the particular total sales. This limit on the one hand and some other critical points of the valid Electricity Feed Law on the other, which led to a constitutional challenge by a utility, are recently arising uncertainties. It becomes obvious by the German development of wind power utilization, how important creation and maintenance of reliable framework conditions are. The design of the in relation to the energy law amended Electricity Feed Law will be addressed in section 3.2.1.

Positive experience with wind power encouraged to start a programme for photovoltaic in 1990. This so called Thousand Roof Programme was financed by federation and states. Applicants could receive an investment subsidy of 70%. To the end of the programme in 1993, 2250 application with a capacity between 1 and 5 kW were financed. The programme was followed up by evaluation and research (Sandtner, 1996). The programme led to technical improvements but could not establish a boom like in wind energy with the consequence of creating a market for important solar cell producers in Germany (Solarthemen 18,1997, p. 37). The number of photovoltaic units used in this programme was too small to lead to important effects in cost reduction. Even in conjunction with the Electricity Feed Law no additional effects were created. As will be shown in 3.2.4 the new model of pricing on cost basis that has been introduced in some cities, gives much better chances to lead to a higher diffusion of photovoltaic systems.

A comparable programme could have had strong effects in the area of biomass. After the introduction of the Electricity Feed Law combined heat and power generation developed only slowly. The drop of constraints regarding energy from biomass of agriculture and forestation as well as landfill gas introduced with the amendment of the Electricity Feed Law can support a stronger development.

3.2. Present Conditions

3.2.1. Wind energy

Regarding wind energy projects in Germany the following laws are relevant: building code, law for natural protection, law for environmental protection and energy law (Forschungsgruppe Windenergie, 1997; Hartmann, 1995; Molly, 1990).

A permission for construction is based on the following aspects static calculation, proof of security, equipment relevant for security, distance to neighbouring lots, compatibility with regional plants, security of aviation, possible magnetic interference, reflections and shadows, deterioration of landscape. In some cases the law for the protection of buildings, the law of roads, the law of water ways and the law for the protection of military areas can be relevant for wind power projects.

On the first of January 1997, wind converters have been privileged in areas outside of human dwelling. However, this does not imply that any number of locations is available in these areas. Some of the criteria mentioned above are still relevant. Local governments can control the development of wind energy through setting up of regional plans until the 31st of December, 1998 in which they show areas available for wind energy. Such plans of course are only meaningful if wind is available, the connection to a nearby grid is possible, and all restrictions from natural and landscape protection have been observed.

Particularly the concerns of natural protection and landscape protection lead to a conflict of interest regarding the already existing density of locations. Special emphasis is on the influence on animal life, particularly seldom breeds of birds. Therefore, in some cases, wind converters only receive permission if additional substitutional measures for landscape protection are being taken. From some areas of natural protection or bird protection the distance has to be at least 200 metres, if a seldom breed of bird is at stake 500 metres.

On the basis of the environmental protection law the minimum distance to dwellings is 200 - 500 metres which is meant to reduce disturbances from noise. These rules can lead to problems for planned wind converters in cities. Often in cities there are initiatives to erect a wind plant, partly to demonstrate environmental consciousness or demonstrate that the future is in renewable energy.

The energy law requires that the electric utility responsible for that region shall be informed about the plant production of electricity. At the moment, wind converters in general are not being controlled, however, it is possible to change this. In some states this is already the case. Schleswig-Holstein has prescribed a security check for private wind energy converters every two years.

The association of power producers (VDEW) has developed technical conditions for connection to the grid. In general regarding the technical aspects of wind energy, the framework conditions are reliable whereas regarding the permission procedures there is a large amount of discretion for the authorities.

Local governments which are important in the administrative procedure for permission often are split-on the one hand the wind converter as an investment will lead to tax income on the other hand they are afraid to lose some qualitative factors of their location through each additional wind power project.

As has already been mentioned, wind power has profited ideally from the electricity feed law. Because all favourable wind power locations are located in coastal areas this leads to a disadvantage for electric utilities in those areas. After the amendment of the Electricity Feed Law the purchase duty relates to the transmission network operators of the particular area, or "das Unternehmen, zu dessen für die Einspeisung geeignetem Netz die kürzeste Entfernung vom Standort der Anlage besteht" (§2). To limit the burden on single utilities, §4 contains a hardship clause: If the share to be paid for according to the Electricity Feed Law exceeds 5% of the total sales of a company, the mother company has to pay for it. If there is no mother company, above the 5% limit, the duty to purchase from additional plants is being dropped. For the utility PreußenElektra this prerequisite will be matched by the turn of the century, if the current volume of new wind power construction continues. Thus, regardless of the pending constitutional challenge against the Electricity Feed Law some doing-up will soon be necessary. To oppose the disapproval of a guaranteed price by unfavourably affected utilities, it would be sensible to allocate the extra burdens to all German utilities. However, such funds might be too bureaucratic or hard to implement in the existing legal framework. The energy sector would prefer a quota-model, which implies a duty to produce or buy a fixed share of electricity from renewables. The wind power sector, however, is not willing to renounce the actual level of guaranteed prices. The Electricity Feed Law contains an appeal, encouraging the utilities to voluntary self-obligations and an increase of the share of electricity generated from renewables, respectively. However, it can already be noticed, that the up-coming competition on the power market leaves little scope for initiative. Dissimilar commitment in this direction can lead to disadvantages in competition too easily.

3.2.2. The use of hydro power

To use hydro power, usually a river has to be dammed, then water will be directed through the generator and then led back into the river further down. This has an effect on the river and its surroundings. In Germany therefore hydro power stations need permission on the basis of the water law. In this administrative procedure generally an environmental impact analysis is required (Rotarius, 1991). Because of the high population density the high intensity of agricultural and forest use and existing nature and landscape protection for unused areas generally in Germany there are only very limited possibilities for new hydro power generation. Because of the relatively low probability of getting new permissions there is great reluctance to engage in planning for new hydro power.

Therefore the technical potential according to Table 2 is only slightly higher than present production. This difference is rather due to reactivating and optimizing existing installations. There is often the danger that with a modernization administrative requirements will be increased. A usual point of dispute is the required amount of minimum water in the old river which reduces the capacity of the hydro power generator. The different states have developed different requirements. Compensatory measures for ecological damage, additional requirements of natural protection, building of staircase for fish etc. can lead to a severe increase of cost in case of a modernization (Hartmann, 1995).

According to the water law, administration can withdraw the right of use if it has not been used for three years. A large number of locations for small plants thus get lost over time.

In addition the water law requires that waste caught in mechanical devices cannot be returned into the river, thus considerable waste disposal costs arise. For large plants the amount of waste collected is so large that it pays to add a plant to burn wood residues for producing heat. In case of modernization or reactivation also requirements of the building code and the energy law have to be considered.

The Electricity Feed Law takes care of the electricity price for installations up to 5 MW. The price for electricity from hydro power is lower than from wind or photovoltaic systems. There is a distinction between small installations below 500 kW receiving in 1998 0.149 DM/kWh (i.e. 80% of average revenue) and larger installations over 500 kW, where the price is between 0.149 and 0.121 DM/kWh (decreasing to 5 MW, lowest limit being 65% of average revenue) depending on peak capacity. In comparison, wind generators receive 0.168 DM/kWh independent of capacity.

Subsidized interest rates or tax allowances often are available. Sometimes there is also the chance to get investment grants from the programme of single states or from the European Union.

3.2.3. Geothermal energy

Geothermal energy is regulated by water law, resource law and the building code (Forum für Zukunftsenergien, 1992; Hartmann, 1995). Accordingly, different administrations (water, resource and construction) have to cooperate. Ground water should not be contaminated and recycling of used thermal water is required to maintain the resource.

Multiple use without recycling could improve the economic perspective. On the other hand thermal water in some regions is salty and cannot be directly led into rivers. For pumping and heat pumps, noise emissions have to be observed as well as the rules for emission reduction in engines.

3.2.4. Photovoltaic

Photovoltaic cells usually are integrated in roofs. The use on facades is seldom and there are few examples of installation directly on land. Except for the general rules for technical safety, no special rules have to be observed. In some states not even a building permission is required, usually it is sufficient to just inform the authorities. The weight of the modules (25 kilogram per square meter is within the tolerance of static norms for roofs so that generally no special proof for the ability of the roof to carry these weights is required (Hartmann, 1995). There could be problems in exceptional cases if buildings which are protected for historical reasons and solar cells appear to be foreign.

Regarding the feeding of electricity into the grid the same rules apply as have been mentioned in section 3.2.1. for wind energy. It is generally recommended that experienced firms plan and install these photovoltaic cells because many details are connected with a number of norms and rules.

Regarding the high investment costs there is no economic incentive for photovoltaic electricity generation. Investment grants and the guaranteed price according to the electricity feed law (1998 16.8 Pfennig/kWh as for windpower) however can move some people to decide for a photovoltaic generation although it is not economic.

Since a short time some utilities practice a price covering cost for private photovoltaic installations. The utility gets the right to increase its electricity price by a small margin so that they can pay about 2 DM/kWh for electricity from photovoltaic installations. This way the tariff customers subsidize photovoltaic installations in 26 cities up to now. The price increase is in the order of magnitude of 0.3 Pfennig per kWh (Solarthemen 17,1997).

3.2.5. Solar thermal

Solar collectors usually are used for the production of hot water and for heating swimming pools. In Germany some project have been realized now where large collector fields and seasonal storage in connection with heat distribution system ('solar district heating') leads to high solar contributions to heating of buildings. Collector system integrated into building a roof pose no big administrative problems. Location on earth that can play a role for district heating sometimes needs permission only if the height exceeds 0.8 meter.

If storage is required for solar district heating the rules for water protection have to be observed. Because of the protection of drinking water, for solar hot water only anti frost chemicals are allowed which pose no health problems.

The pressure in the systems has to be controlled by safety valves on the basis of the rules valid for steam production. The collectors have to be certified by type or individually. In addition, special technical norms have to be fulfilled where the DIN 4757 that is directly related to solar thermal installations at the moment is being adjusted to international norms.

In Berlin it was tried to issue a directive which required solar collectors generally for new buildings. This however was not realized. For solar hot water federation, states, local government and utility companies offer investment grants (Solarthemen 20,1997, p. 5). Usually they are related to collector size. Often finance offered is used up in a short time or the grants are relatively low. Other programmes are of a more long term nature and offer low interest loans or grants in the framework of ecological measures regarding new buildings.

3.2.6. Solid Biomass

As has been shown in section one, the energetic use of biomass can take many forms. In this section we deal with the most important form which is direct burning of residues from wood or straw. There are heaters for single family buildings as well as large installations for the production of heat or the combined production of heat and power for district heating systems. According to size and number of peripheral installations required, a different number of administration has to deal with the matter.

The most important rules are those regarding emissions. (Federal law for the protection of the environment and the directives on that law as well as so called technical annex air) (Hartmann, 1995). The standards for emissions have been reduced continuously so that technical possibilities are being expressed in those norms. Traditionally, heaters for biomass had the disadvantage compared to oil or gas that there were high emissions of dust and a high share of incompletely burned carbohydrates (high CO content).

Through design of the burner, control of air and filter techniques on the side of the flue gas enormous improvements could be made and partly also increased the efficiency of the burners. Table 4 gives a survey of emission standards depending on the capacity of the installation.

For burning of straw, installations for burning of straw have to fulfil high requirements that are given for untreated wood only above 1 MW. If contaminated wood is used, standards are higher and administrative procedure for permission is more complex.

There are only few examples of combined heat and power production. Partly these are in the wood processing industry. Generally, these are based on steam turbines and are economical because fuel in this case is free. New developments are with gasification of wood in connection with a motor driven generator. Such installations are able to use contaminated wood residues (Schulz, 1997). In this area the electricity feed law is not relevant because either, because the own use of electricity leads to a higher credit or the credit for waste disposal is more relevant. For co-generation plants of up to 5 MW on the basis of solid biomass a guaranteed price according to the Electricity Feed Law is granted (1998: 14.9 pf./kWh).

In small units only the burning of natural wood is allowed. It would be justified to allow also the burning of briquettes from sawdust, straw or wood chips (splints) because these do not require a sophisticated technique for burning. A contradiction in the existing rules is that emission measurement for small units is only done once at the beginning if the feeding is done by hand. Installations fed mechanically that usually have much lower level of emissions of the size of 15 kW have to be measured every year (Hartmann, 1995).

The problem with bio fuel is that the quality of burning changes with load. For plants with manual feeding at more than 15 kW now operation at full load has been prescribed if emission standards cannot be maintained otherwise. This requires a large storage. Another problem is the changing quality of the fuel which can lead to emission problems also in good plants. A classification has up to now only taken place for wood briquettes.

In addition to emission problems, building code and fire safety are important. In addition the rules for the protection of labour and general security rules have to be observed. Sometimes also permissions on the basis of water law are relevant. Manufacturer and planner have to observe norms and many details.

Concluding, we can say that the rules for using renewable technologies are not more far reaching than those for conventional technologies with similar potential dangers. There are quite a number of technically sufficient solutions on the market that fulfil the requirements. However, except for the federal state of Bavaria there is a lack of reliable programmes of promotion. There are incentives by tax allowances and sometimes investment grants.

Table 4 Emission standards for biomass[3)]

Capacity	Oxygen O_2 [Vol.%]	Emission standards				
		CO [g/m^3_n][1)]	dust [mg/m^3_n]	Total C[2)] [mg/m^3_n]	NO_x[3)] [mg/m^3_n]	SO_2 [g/m^3_n]
Emission standards for burning of untreated wood:						
15-50 kW	13	4	150	-	-	
50-150 kW	13	2	150	-	-	
150-1000 kW	13	1	150	-	-	
500-1000 kW	13	0.5	150	-	-	
1-5 MW	11	0.25	150	50	500	2.0
5-50 MW	11	0.25	50	50	500	2.0
Emission standards for burning of straw and similar biomass:						
15-100 kW	13	4	150	-	-	-
0.1-5 MW	11	0.25	150	50	500	2.0
5-50 MW	11	0.25	50	50	500	2.0

[1)] m^3_n = normal cubic meter at 0°C and 1013 mbar
[2)] volatile organic compounds are given as total C
[3)] given as NO_2
Source: Hartmann, 1995, p. 345.

3.2.7. Anaerobic processes

Production of biogas from anaerobic digestion takes place in the treatment of sewage water and becomes more and more important for the treatment of commercial waste water, the treatment of biogeneous parts of domestic waste and animal waste in agriculture. Anaerobic processes in garbage deposits also lead to the production of gas which is comparable to biogas (landfill gas). In this case only gas collectors have to be integrated into the garbage system whereas in all other cases heated reactors, different according to the type of waste, have to be built. Because of the many different applications and the necessary treatment of residues that come out of this process, a considerable amount of legal rules have to be observed (Schachtner, 1993; Hartmann, 1995) of which we can describe here only the most important ones.

In 1996 the law of recycling and waste came into force. This law prescribes energetic or material recycling and in conjunction with the technical annex for domestic waste from the year 2005 it will not be allowed to deposit untreated biogeneous waste in waste deposits. On the basis of this law new directives have been issued that go into details regarding different types of waste.

In addition by introducing the Electricity Feed Law guaranteeing a minimum price this became a matter of interest. However, the prices granted are restricted to electricity from waste treatment or landfill gas between 0.121 to 0.149 DM/kWh[4] (compared to 0.168 DM/kWh for wind power) and is mandatory for installation below 5 MW_e. Because sewage plants have a high internal need for electricity and therefore sell only small surplus quantities it is not understandable why also in these cases a capacity limit has been prescribed in the law. Regarding agricultural biogas plants the minimum price up to 5 MW_e at present is 0.149 DM/kWh independent of capacity. Commercial waste had been excluded before the amendment in 1998.

In many agricultural biogas plants the treatment of commercial waste or organic waste from households leads to improvements of the economic situation. In addition to further credits to the operator the concept leads to safeguarding the agricultural use of residues.

Plants for anaerobic digestion are subject to building code, water law, environmental protection law and sometimes energy law. The building code is important because only new buildings without heaters up to 20 cubic meters of space and containers of up to 5 m^3 space do not need permission and even the changed use of existing buildings can lead to the necessity of receiving a permit.

General commercial rules apply to safety aspects for instance the directive on pressure containers regarding the digester or gas store if the pressure is over 0.1 bar. Also the directive on electric appliances in rooms with the danger of explosion and others have to be observed.

The federal environmental law with all directives is not very relevant so that usually technique available on the market can be used. Because of dangerous elements in a gas in the case of landfill gas there are special requirements which however do not require gas purification. For heaters based on biogas an administrative procedure is required from 10 MW and for motor driven co-generators from 1 MW capacity corresponding to about 300 kW_e. Small plants very often fulfil high emission standards to correspond to conditions of promoting agencies.

Generally it is no problem to stay within the allowed emission limit. For some materials however, it may be necessary to desulphurize the biogas which at the same time offers higher protection against corrosion. In addition, standards of maximum noise emission have to be observed in the use of the gas.

A constraint for an optimal use of gas was, that a concession only was issued to the area utility, this has changed with the amendment of the energy law in April 1998. Now there are only minor obstacles for operators of a biogas plant to supply third parties with biogas or electricity. (Admission as a utility, concession contract for using public ground for lines etc.)

Agricultural use of residues from gasification is subject to the directive on residues from waste processing which defines a standard for harmful substances and maximum amounts. In addition the directive on fertilizer use as part of the fertilizer law has to be observed. There are prescriptions regarding the amount of fertilizer to be used in certain cases.

Some states give financial subsidies as well as some local governments. Normally subsidies are limited and only given in special cases.

3.2.8. Biogeneous liquid fuels

The production of bio diesel on the basis of rape has increased since the agrarian reform of 1992. This reform was meant to control the overproduction of food and the corresponding fall in production prices. Farmers have to set-aside 15% of their agricultural land and are compensated with premiums for set-aside land. These premiums are also available if the land is used for energy plantations. Thereby the production of bio diesel or ethanol is being subsidized. One litre of bio diesel from plants (methylester from rape) is, at the moment, about 5% cheaper than normal diesel, which however, has a somewhat higher energy content.

Energetic use of bio ethanol is achieved by mixing it with gasoline, according to an EU directive, up to 5% being allowed. The addition of ethanol improves the properties of the fuel by increasing the beat consistency.

Emission standards are important for the use of bio fuels. There has to be a certification on a type basis on which it is being decided whether a special engine is certified for bio diesel.

4. Renewable energy in the future market order

As mentioned in section 3.2.1., the Electricity Feed Law being the most important promotion instrument, lacks the prerequisites of being supported by an overall consensus and of being designed suffiencently long-term. Thus alternative protection measures have to be considered regarding the future structure of the energy sector to be expected:

The electricity sector in Germany is in a phase of transformation from the closed model of vertically integrated enterprises in the direction of an open model with elements of vertical disintegration (separation of transmission and production) and the development of specific markets for power and transport services.

The development at the moment cannot be anticipated clearly because the legal reform is still in the process. Such changes need some time. In an open market electricity can be bought and sold according to the time orientation of the contract (contract market and spot market). The working of such markets strongly depends on the rules for negotiated access and transmission pricing.

Power production from renewable energy will only have a perspective if it gets priority or a financial bonus. In addition to an energy tax for electricity production from fossil fuels or nuclear energy which would automatically improve the economy of renewable energy, two possibilities are given:

(1) legal fixation of electricity price

The electricity price to be paid for electricity from renewable energy will be fixed by considering an environmental bonus. Additional costs can be paid out of a fund which will be fed from conventional power production. The disadvantage of this system is the danger of over-subsidy. The bonus is connected to the technology but not to the economy. Thus also plants could receive the bonus that do not need any subsidy. Such a subsidy could also lead to higher prices for plants which would be economically inefficient.

(2) quota for power from renewable energy.

Electricity suppliers have to buy a certain quota of their power from renewable energy. The quota is being fixed by the state. In this context it does not matter if the plants are being operated independently or by others. This model is comparable to the trade with environmental certificates. The duty to buy such 'green' power being the same for all, such duty is neutral to competition. Suppliers of electricity have an interest to keep the cost for power from renewable energy as low as possible. Therefore, in such a system, a market for power from renewables can arise if for the electricity sector in general the conditions of an open market prevail. The great advantage of such a solution is that government acts only as a regulator and does not have to move any financial means. At the same time, enterprises have a clear basis for planning if the quotas for power from renewable energy are fixed on a long term basis.

References

Becker, H. (1997), *Die Aufstellungszahlen der Hersteller für das erste Halbjahr liegen vor: Die Spitze setzt sich weiter ab.* in 'Wind Energie Aktuell' 9.

Diekmann, J., Horn, M., Hrubesch, P., Praetorius, B., Wittke, F. and Ziesing, H.-J. (1995), *Fossile Energieträger und erneuerbare Energiequellen*, ed. Forschungszentrum Jülich GmbH (KFA), Serie: Monographien Instrumente für Klimagas-Reduktionsstrategien (IKARUS), Jülich.

Durstewitz, M., Hoppe-Kilpper, M., Schmid, J., Stump, N., Li, T. and Windheim, R. (1996), *The German 250 MW Wind Programme - A Success Story*, ed. Deutsche Gesellschaft für Sonnenenergie e.V. - DGS; EuroSun'96 10. Internationales Sonnenforum Proceedings, DGS-Sonnenenergie Verlags-GmbH, München.

Forum für Zukunftsenergien (1992), 'Erneuerbare Energien - ein Leitfaden für Städte und Gemeinden'.

Forum für Zukunftsenergien (1997), 'Förderfibel Energie', ed. Fachinformationszentrum Karlsruhe, Verlag Deutscher Wirtschaftsdienst.

Hartmann H. and Strehler, A. (1995), 'Die Stellung der Biomasse', ed. Bundesministerium für Ernährung, Landwirtschaft und Forsten: Schriftenreihe Nachwachsende Rohstoffe Bd. 3, Landwirtschaftsverlag, Münster.

Internationales Wirtschaftsforum Regenerative Energien (IWR), Forschungsgruppe Windenergie, 'Grundsätze für die Planung und Genehmigung von Windenergieanlagen'. (Internet-Info: http://www.uni-muenster.de/Energie/wind/raum/ro-prog/leit-nw2.html)

Johnsen, B. (1998), *Wind/Energie/Aktuell – Rekordjahr in Wartestellung*, in Erneuerbare Energien.

Langniß, O., Nitsch, J. and Trieb, F. (1997), 'Vorschlag für ein Sonderprogramm zur beschleunigten Markteinführung regenerativer Energien bis 2010', unveröffentlichte Expertise für die *Gruppe Energie 2010*, Mai.

Leuchtner, J. and Prüser, K. (1994), 'Photovoltaik - Anlagen - Marktübersicht 1994/95', ed. Öko-Institut e.V., Freiburg.

Marko, A. and Braun P. (1997), 'Thermische Solarenergienutzung an Gebäuden', Fraunhofer Institut Solare Energiesysteme, Springer-Verlag, Berlin.

Meissner, D. (1993), 'Solarzellen, Physikalische Grundlagen und Anwendungen der Photovoltaik', Vieweg-Verlag.

Molly, J.-P. (1990), 'Windenergie - Theorie, Anwendung, Messung', C.F. Müller-Verlag, Karlsruhe.

Rotarius, T. (ed.) (1991), 'Wasserkraft nutzen - Ratgeber für Technik und Praxis', Cölbe

Sandtner, W. (1996), 'Ergebnisse des Bund-Länder-1000-Dächer-Photovoltaik-Programms' in *Elektrizitätswirtschaft*, vol. 1, pp. 35 ff.

Schachtner, K. and Uszynski, E. (1993), 'Rechtliche und technische Vorschriften für Bioabfälle', ed. Rheinisches Institut für Ökologie, Köln (RHNO).

Schüle, R., Ufheil, M. and Neumann, C. (1997), 'Thermische Solaranlagen Marktübersicht', ed. Öko-Institut e.V., Freiburg.

Schulz, W. (1997), 'Stand der Holzvergasungstechnik', ed. Bremer Energie-Institut, Bremen.

7 Development of Energy and Environmental Regulation in Lithuania

REMIGIJUS CIEGIS AND VIDMANTAS JANKAUSKAS

1. Introduction

In the centrally planning and control economy, as it was in Lithuania almost for fifty years, regulation of the energy sector was performed by the Government, one may remember that all utilities were state owned. Environmental issues were highly neglected, so energy sector was making significant negative impact on the surrounding environment. Power plants were burning high sulphur heavy fuel oil and no cleaning equipment was installed at any plant, neither at the refinery where this fuel was produced. Lakes at the power plants used as coolers were overheated (this caused especially serious problems for the living species at the Visaginas lake used as a cooler of the Ignalina Nuclear Power Plant).

Lithuania, like other former communist countries, is facing many difficulties in restructuring from a centrally planned economy to a market related one. The move to commercialisation, liberalisation and introduction of private finance in the energy sector has led to a reconsideration of the appropriate industry structure and also a move to a new type of regulation. With decentralisation and liberalisation of the national economy the main goal of the energy policy should be: preparation of laws, setting of the priorities and control mechanisms. Energy sector regulation and supervision, management of state owned enterprises, research and development, new plants and design should be transferred to independent regulatory institutions and the private sector. The main functions of the independent regulatory institutions should be the protection of energy consumers interests and rights, implementation of consistent pricing system and the regulation of natural monopolies.

This paper describes changes in the energy sector structure, analyses development of energy demand and supply, the main part of the paper is devoted to the description of the energy and environmental regulation development: regulation institutions, regulation tools, and implementation of the international experience.

2. Energy supply

The technical and institutional structure of the energy sector, inherited from the past does not suit the requirement of Lithuania as an independent country and the conditions of a market economy. It is necessary to restructure and improve the whole energy sector to the totally new conditions.

2.1 Electric power supply system

The Lithuanian Power System covers the entire area of the country. It was developed as part of the former Soviet Union's north-western interconnected power system which integrated the Lithuanian, Latvian, Estonian, Belorussian, Karelian, Kola, Leningrad and Kaliningrad power systems. After 1991, the Lithuanian Power System operates with the Latvian and Estonian power systems. All the power systems are controlled from the Baltic Dispatch Centre in Riga. At the end of the Soviet regime, the three Baltic countries produced 52 TWh of electricity, consumed only 37 and exported the surplus to Russia and Belarus. The Lithuanian Power System includes the Ignalina Nuclear Power Plant, Lithuanian Power Plant, the cogeneration plants (CHP) of Vilnius, Kaunas and Mazeikiai, the Kruonis Hydro Pump Storage Power Plant, Kaunas Hydro Power Plant and regional district heating utilities.

Assets of the Ignalina NPP comprise about a half of the total assets in the energy sector. The importance of the plant is evident during the current period of economic reforms. Due to lower production cost and higher reliability of fuel supply the Ignalina NPP has become the main producer of cheap energy. This plant alone is able to produce low cost electricity and guarantee electricity export in future. Moreover, only if it does exist the Kruonis PSPP is economically efficient. The existence of the Nuclear Power Plant assures the reduction of the atmospheric pollution.

Safety of operation is the most critical problem now. Under technical assistance of international institutions the operator is devising various measures to improve safety of the Ignalina NPP. All the safety upgrade measures, spent fuel storage included, institutional restructuring should be discussed with international institutions.

2.2 Natural gas and oil supply in the country

The length of the main pipelines in Lithuania is more than 1400 km but a large number of potential customers are not connected to the grid. In the National Energy Strategy it is suggested that an expansion of the distribution system should take place over the next ten years. Depending on the heating system being used at the moment new customers could be connected as direct gas users or via their district heating systems.

The future role of gas in Lithuania will be highly influenced by the fate of the Ignalina Nuclear Power Plant as also of the development of oil industry. If the Ignalina NPP were to be shut down it is likely that a gas powered generation station would be considered due to the rapid construction programme possible. It would increase natural gas demand significantly. The other important factor: presence of (possibly cheaper) a heavy oil in a local market together with low environmental requirements would hinder expansion of a gas use, especially in the power generation.

The main supplier of petroleum products in the country has been the Mazeikiai Refinery 'Nafta'. The total capacity of the refinery is some 12 million tonnes of crude oil annually, this being reached by a step by step expansion over the last decade. All the crude oil refined at the Mazeikiai Refinery was imported from the CIS by pipeline. There are two domestic pumping stations, at Birzai and Joniskis. From Birzai the pipeline has two branches one going to Mazeikiai and the other to the Ventspils crude oil export terminal in Latvia. The refinery is one of the most modern in Eastern Europe and it could be upgraded to produce products that are in demand in Europe. The refinery produces at the moment large volumes of high sulphur heavy fuel oil. The export price for this is low and its domestic use causes pollution. Investments needed in upgrading of the refinery will be coming soon as the strategic investor Williams International has bought the stake at oil sector.

There is a modest but significant potential for Lithuania to produce its onshore oil of some 50 million tonnes of recoverable reserves. Current production (1996) was only 160 thousand tonnes but is gradually increasing and according to various sources may reach 1 million tonnes per annum. Offshore reserves are also potentially significant but exploitation will depend on resolution of state boundaries disputes in the Baltic sea.

2.3 Indigenous renewable energy resources

Share of indigenous energy resources is rather small in the energy balance for Lithuania - only 3-4% of total and only during the last several years increased to 6-8%. During the last several years energy production from wood, peat, hydro and other indigenous resources (except of oil, this production was increasing) has remained almost constant. The contribution of these energy sources to overall energy consumption in the country has, however, increased because the total fuel consumption has dropped.

A rough estimate of a technically harnessable energy potential from new and renewable resources suggests that a maximum of about 15% of the energy demand in Lithuania could be covered in the future by indigenous energy resources. The figure of 15% as a long-term target is perhaps over ambitious, given the costs and the changes that would have to take place to achieve that goal.

Peat is one of the indigenous resources with the largest theoretical potential of the resources in Lithuania, and of the leading resources, the one with the most straight-forward method of collecting and using the resource (when compared with the other resources with a large theoretical potential, such as solar energy or straw). The 270 million tonnes of peat estimated as the total potential, but at present only about 100 thousand tonnes is annually used. Milled peat is presently used in a few boiler houses of different peat enterprises.

The future potential for energy production from wood is estimated to be 0.5 Mtoe, 80% of it is assumed to be used to cover the wood industries own energy demand and as fuel privately used in small individual boilers, the remaining share would be available for energy production in district heating systems.

Municipal waste generated in the largest cities constitute a realistic energy potential. The municipal waste amounts generated in these cities are estimated to be more than 30 thousand tonnes annually which corresponds to a potential energy production of 50 ktoe per year. Presently such wastes are usually disposed of in municipal landfills. Disposal at landfills leads to a large number of environmental problems, particularly as some landfills lack adequate control and protection measures. The Government of Lithuania recently issued decisions meant to improve the existing municipal waste management system. These improvements include the prospect of developing some form of energy recovery from the combustible portion of municipal wastes.

There are three sources of bio-gas that can be considered as reliable in Lithuania: those from agriculture (manure of poultry and farm animals), from food-processing (organic wastes), and from municipal waste water treatment plants (with their methane thanks for anaerobic digestion). Only the bio-gas potential in the farms of joint stock companies (i.e. about 300 million m^3 per year), together with the potential from waste water treatment and food-processing industries is considered as relevant as an energy potential.

There are almost 400 rivers longer than 20 km in Lithuania. The combined energy potential of the large and medium rivers only gives approximately 4.5 TWh. Natural conditions in Lithuania are not, in general, favourable for hydro power exploitation, as there are no deep valley where high dams can be built. Moreover, the volume of flow in the rivers fluctuates considerably over the course of the year. There are, however, many smaller existing water storage's throughout the country that could accept small hydro power installations with reduced civil works requirements.

In the Western part of Lithuania geothermal water resources were found. A number of its features, such as high salinity, relatively low temperature and the large investment costs needed to use the geothermal water reduce the otherwise attractive aspects of using this resource. Scientists of the Geological Service of Lithuania have estimated the technically harnessable energy potential to be some 8 TWh/year, but only about 10% of it is available for energy production.

The coastal area is the region with the highest wind energy potential. The average wind speed there is 5 m/s at 10 m height. As regions with wind speeds less than 4 m/s are not recommended for wind projects most of the non-coastal Lithuania is not suitable for wind energy sites.

With a growing awareness in the country that indigenous energy resource could play a larger role in satisfying energy demand the Lithuanian Government has introduced a number of pieces of legislation. One of the main energy policy objectives stated in the Energy Law is promotion of domestic and renewable energy resources.

3. Development of energy and environmental regulation

3.1 Institutional development

In June 1995 all the Baltic states signed the European treaty and became associated members of the European Union. This requires all the three states to restructure their economies in line with the European requirements. For the energy sector it means market liberalisation, transparency of pricing, security of supply, application of certain safety and environmental standards. All new legislation drafted in these countries is adapted to the EU requirements.

Energy Law was passed in 1995 and the Nuclear Energy Law was passed in 1996. The Ministry of Economy (it has two departments responsible for the energy sector supervision), according to the Energy Law, regulates energy activities representing interests of the state. It implements the state energy policy objectives and acts according to the regulations confirmed by the Government. Other energy regulatory institutions are: the State Energy Inspection, the State Nuclear Power Safety Inspectorate, the National Control Commission for Prices and Energy.

There is the State Nuclear Power Safety Inspectorate in charge to ensure that the operation of the Ignalina NPP is in full compliance with the requirements imposed by the Lithuanian regulations. Nuclear Safety Inspectorate is an independent agency, its head is appointed by the Prime Minister.

Members of the National Control Commission for Prices and Energy are proposed by the Government and appointed by the President of Lithuania for the five year period. The main objectives of the Commission are: to set energy pricing principles and to implement energy policy goals in the control of energy activities. The Commission analyses the main energy economy issues, determines tariff calculation principles, approves the tariff calculations performed by the regulated utilities and monitors implementation of these tariffs.

The basic objective of the State Energy Inspectorate is to perform the state supervision over energy equipment in order to ensure reliable, efficient and safe supply and utilisation of energy resources. The other sphere of involvement for the Inspection is the licensing of energy activities - transportation and distribution of oil and gas, production and transfer of heat and electricity, operation of energy equipment.

3.2 Development of the legal framework

Energy Law is the main law regulating activities in the energy sector, it was passed by the Parliament in 1995. It sets down broad principles only and gives the general legal framework for the energy sector development.

The Law explains the aims of the National Energy Strategy, its structure, its timing (the strategy was developed for a 20 year period, but should be revised every five year period), and its role in the energy sector development. It defines the following main objectives of the Lithuanian energy policy :

- energy saving and energy efficiency;
- diversification of fuel supply by sources and by countries;
- stimulation to use renewable and secondary energy resources;
- reduction of the harmful effect of the energy sector on the environment;
- encouraging of competition and participation of private capital to increase economic efficiency, etc.

The Privatisation Law was passed by the Parliament in March 1991. It set the rules for the primary privatisation of the state owned assets. Currently, the privatisation process is implemented by the Central Privatisation Commission and its executive arm Privatisation Agency. There are two ways of privatisation available: auctions and selling of shares. Foreigners are allowed to buy into privatised assets if they are included in the special list of assets to be sold for hard currency.

Nuclear Energy Law, passed at the end of 1996, regulates public relations arising out of the use of nuclear power for generation of electricity and heat. The law provides a legal basis for natural and legal persons to operate within the sphere of nuclear energy production. The purpose of this law is to ensure nuclear safety and the use of nuclear energy for satisfaction of peaceful needs only.

The Environment Protection Law was passed by the Parliament in January, 1992. The law deals, among others, with :

- monitoring of natural resources use;
- monitoring of the state of environment and the negative impact on environmental systems;
- regulation of economic activities;
- economic mechanism for environment protection, a - o.

In 1996 amendments to the Law were passed by the Parliament, all the articles were revised and rearranged according to contemporary requirements.

The Law on Royalties (taxes on natural resources) was passed in 1991 and amended in 1995. The purpose of this law was to increase responsibility of natural resource users, and to encourage efficient and economic utilisation of the national wealth at their disposal. Revenue accumulated by these taxes serves to offset the state expenditures needed to monitor the use and state of natural resources.

The Law on Pollution Taxation was passed in 1991 and amended in 1995. This law imposes mandatory fees for all persons, businesses or organisations whose activities contribute to the pollution of environment, there by implementing the 'polluter pays' principle.

3.3 Stimulation of the renewable energy

Despite the fact that Lithuanian energy sector is overburdened with excess capacities, promotion of renewables is one of the main energy policy objectives. Renewables may help to solve at least two problems: increase diversification of the energy supply and reduce harmful pollution of the environment.

Renewable energy is one of the priority directions of the National Energy Efficiency Programme, revised in 1996. Utilisation of renewable and other indigenous energy resources was stimulated by various decrees and decisions of the Government. In 1993 the Government decision 'On better use of indigenous fuels' was approved, which encouraged the preparation of proposals for the production and modernisation of indigenous fuel combustion boilers (especially, wood chips). It made tax reductions available to try and encourage better utilisation of indigenous fuels. Tax reductions were revised in 1995 by the Governmental decision 'On reduced profit taxes for the priority industries of Lithuania'. This decision reduced the profit tax from 29% to 15% for some industries involved with the production of indigenous fuels, and the manufacture of equipment necessary to prepare and burn such fuels.

Governmental decision 'On agricultural hydro engineering units use for small hydro energy installations' (1995) promotes the installation of small hydro power plants. This decision allows the construction of small hydro power plants at existing water reservoirs owned by the state. Foreign investors are also allowed to participate in these projects.

Use of renewable energy resources is also regulated by the Energy Law. The law obliges the excess energy produced by autonomous producers to be sold into the national grid. The price of electricity bought from an autonomous producer is determined by an agreement with the Ministry of Economy and normally equals to the current average tariffs.

4. Restructuring of the energy sector and its regulation

4.1 The current structure

In the Lithuanian energy sector still prevail highly vertically integrated monopolies, but restructuring of the sector goes on.

Lithuanian Power Company (LPC) is a highly centralised, vertically integrated monopoly structure, apart from the power generators it also owns the cogeneration plants. With the recent (1997) decentralisation of the Lithuanian Power Company besides the Ignalina Nuclear Power Plant, which is a separate fully government owned company, there appeared more participants in the electricity sector which are not a part of this company. Lithuanian Power Company buys power from the Ignalina NPP and cogeneration plants and resells it to final customers.

The gas sector is currently dominated by one vertically integrated company Lithuanian Gas. This company purchases gas on the state border from the Russian gas company Gazprom and supplies it to the final customers. Private company Stella Vitae also buys gas on the state border and resells it to the Lithuanian Gas. Different affiliates of the Lithuanian Gas company cover the production of gas equipment, transportation of gas via pipelines, operation of the gas systems (including the LPG marketing) and gas distribution.

Lithuanian petroleum industry was vertically separated and deregulated. The refinery, the oil pipeline, the oil harbours and the distribution network were operated by different companies. But in order to attract foreign investments into the oil sector the refinery, the oil harbour and the oil pipeline were merged into one company and one third of this company's shares were sold to the strategic investor Williams International.

4.2 The electricity sector

Two combined heat and power plants in Vilnius and Kaunas in 1997 were separated from the Lithuanian Power Company and transferred to municipalities. Restructuring of the Lithuanian electricity sector was proposed in the recent Report by *Price Waterhouse*. This Report highlighted the main problems within the electricity sector:

The main issues of the power sector determined by the authors of that Report were as follows:

- lack of competition and commercialization in any part of the sector,
- lack of the consumer focus,

- privatization is not going, there are many barriers for it.

In general, the PHARE experts conclude, institutional structure and management of the Lithuanian electricity sector are unsound and reforms in the sector are necessary.

It was proposed in the first stage of reforms to separate generation from the transmission network, and the last one at the beginning to keep together with the distribution network. Separation of generation does not necessarily mean that power plants will be privatized. At the beginning all electricity generation may be connected into one business unit, having its own budget, account, independent management and transparent relations with the other technological units.

Experts propose, that in future every power plant may become an independent unit (this proposal was sharply criticised by the LPC management, as in the West electricity generation companies ussually consist of several plants). It was also proposed to transfer cogeneration plants to municipalities and the pumped storage hydro plant was left controlled and managed by the central dispatch.

Generation is a part of the electricity sector where competition is possible and should be promoted. Pure competition in generation, as in England or Scandinavian countries, in Lithuania is hardly possible, not because Lithuania is a small country, but because the exiting units of the power plants are very large – there is not enough of the same weight of market players. The PHARE experts propose to establish the Power procurer, it would purchase electricity produced at the power plants using non-discriminatory contracts with producers. Such a model is applied in Portugal, Northern Ireland, Hungary, it is one of two models allowed by the EU Electricity Directive. Energy procurer make contracts with the producers, regulating by these contracts investments, operation costs, reserve payment, payment for the electricity supplied and service standards. It is proposed in the Report to separate Power Procurer from LPC and transfer its control to the Minister for Energy.

Electricity transmission by the high voltage lines and central dispatch are functions of a natural monopoly, they should be controlled by the LPC. In the first stage, as it was mentioned, there was no proposal to separate distribution from transmission, as there was no economic or efficiency justification for the unbundling. Similarly, it was no proposal to establish separate electricity supply companies, their business would be electricity trade, but in the long term companies of this type may be important and useful. There was no support to establish regional electricity and district heat distribution companies, though such proposals were rather popular among some Lithuanian experts and politicians. Only in the later restructuring stages it was proposed to separate transmission from distribution and establish several regional distribution companies, but better less than the current 7 affiliates. An optimal number of distribution companies may be determined based on several criterias:

- guaranties that only financially and technically viable companies are established;
- evaluation of additional cost required for operation of the separate companies;
- capacities of the regulator to compare activities of different companies;
- if distribution companies are not planned to be privatized they may be split into smaller ones.

Therefore, there is a need to have a long term restructuring plan for the whole electricity sector, not only for the LPC. Outline for this plan should be presented as a part of the Plan of Action following the National Energy Strategy. This Plan of Action should be developed in 1999 and approved by the Government. The principal objectives are given by the EU Electricity Directive, the Government should only decide what model to chose: the third party access or the single buyer. As all the three Baltic states decided in principle to establish a common electricity market, this option should prevail in the future plans. In the first stage it is necessary to unbundle electricity generation, transmission and distribution, at least accounts, and transmission should regain some status of independence (independent management). It is necessary for the further reforms, as only independent and transparent transmission service, supplied to all consumers and producers on the non-discriminatory basis, would promote and strengthen market relations in the electricity sector. The Government should approve a plan of action for the market opening: to define clear rules and order, enabling the eligible customers, i. e. customers fulfilling the defined conditions, to purchase electricity directly from the producers; these requirements would be lessened in the future. The best approach is to define the eligible customers not on the technical their interconnection capabilities or specific activities, but on their volume of consumption, e.g. consuming more than 15 GWh annually. Some producers (using renewable energy, combined heat and power plants) may be granted some advantages, defined in the Law or by the Government.

The further reforms in the Lithuanian electricity sector should go in line with the similar reforms in the other Baltic states in order to enter the EU cosequent institutional and technical structures simultaneously. This requires technical, economic, legal and political decisions. From a technical point of view new electricity transmission lines are necessary, they will connect Lithuanian electricity grid with the Polish one and Estonian grid with the Finnish one, these projects could be implemented during the next 3-5 years. Politicians will decide if the Baltic countries will stay firmly connected with the Russian United Power System or will leave it and connect directly to the West European power system UCPTE, or finally there will be no barriers between these two major systems and all Europe, from Gibraltar to Urals will be covered by the united electricity grid.

It is hard to imagine a competitive electricity market in Lithuania until such giants as the Ignalina NPP (even with the closure of one unit at this plant, as it is planned in the national energy strategy, it will be able to meet more than a half of a total demand in the country) and the Lithuanian PP, but common Baltic electricity market is a real feature even in the short term perspective. The present Baltic Dispatch Centre might become an independent system operator responsible for the regulation of power interchanges among the Baltic states as also with the neighbouring countries and regional systems (such as NORDEL). Efficient operation of this agency requires market rules valid in all the three countries as also compatible laws and regulations. If the Baltic electricity market is organised using the Nordic market as an example (Scandinavian electricity market in opinion of many experts is more effective than the British one) bilateral contracts will prevail and in a spot market there will be reserve, frequency stability and other ancillary services traded.

So, a decade from now in the liberalised Lithuanian electricity market large consumers will be able to purchase electricity directly from Latvian, Estonian or even Norwegian producers. One may predict, that the Ignalina NPP operated at a half of its capacity and requiring significant investments in the necessary safety measures as also for its decommissioning, will be hardly competitive in this market. The Government of Lithuania may be forced to introduce a special nuclear levy as it was done in England and Wales in early nineties in order to improve the financial viability of their nuclear power plants. But electricity generation costs of Polish or Estonian companies will be not lower than that of the Lithuanian ones as these countries burning coal and oil shale at their plants will be investing heavily in pollution reduction technologies in order to meet rather strong EU requirements.

4.3 Gas sector

The joint stock company 'Lithuanian Gas' (LGC) is vertically integrated gas monopoly, purchases natural gas on the state border, transports it by the main and distribution pipes (all of them, except of small places, are owned by LGC), supply gas and other services to the customers. Besides that it produces some gas equipment, distributes and sells liquified petroleum gas (LPG) to the consumers. During the last years new companies have emerged, they buy gas from Russia and sell it to the final uses transporting it through the LGC pipelines (the third party access).

In the 'Baltic Gas Study' developed under the EU PHARE programme in 1996 it was proposed to restructure this monopoly. At the first step the non-core activities, production of the gas equipment and LPG trade, should be separated. No one of these activities are related with the natural gas business, therefore may be separated and independent companies established. In 1999 the Government should finally approve the non-core activities separation plan and present the corresponding law on the restructuring of the company to the Parliament.

In the above mentioned study it was proposed to separate distribution from the transmission (transport) and establish legally and commercially independent companies. The reform would be executed in two stages. In the first stage the distribution companies would be owned and controlled by the LGC, but they would act as commercial units. During the transitional period these companies would be commercialized and prepared for privatization. In the second stage the regional distribution companies would become the joint stock companies owned by municipalities or by private investor.

The transmission network (main pipeline) company would be responsible for:

- natural gas import, transport, dispatch and sale to the regional distribution companies (long-term contracts) and large industrial consumers;
- gas storage;
- operation and maintenance of the transmission network;
- planning gas purchase and transmission.
- distribution companies would be responsible for:

- gas sales to the final consumers:
- operation and maintenance of the distribution network;
- planning gas sales.

Technical restructuring plans should be related with the cooperation of the Baltic States and with possible future gas market. Natural gas companies in Latvia and Estonia are already privatized, prices to the large consumers are liberized, the third party access is allowed in Lithuania also. Common gas market may start its operation in the Baltic States soon: gas could be stored in the Latvian underground storage and gas could be purchassed on the Latvian border or elsewhere. In spring,1999 LGC together with the Lithuanian experts started to prepare a new plan for the further reforms in the gas sector, it envisages separation of the distribution and establishing of 3-5 independent distribution companies.

Development of the Lithuanian gas sector will heavily depend on the pace of implementation of the EU environmental requirements. At present, high sulphur heavy fuel oil is much cheaper than natural gas and is pushing away this environmentally friendly fuel from the market wherever competition between these two fuels is possible. If environmental standards were improved, air pollution taxes increased, the situation could become totally different and natural gas may conquer some bigger share of the market. It would encourage technical development of the gas system and its liberalization. With the implementation of the EU Gas Directive gas transport through the main pipelines will be separated from its distribution (at least accounts will be unbundled), it will allow the third party access and eligible customers will be able to purchase gas from other suppliers than the LGC.

5. Environmental regulation

5.1 Environmental impacts

Significant reduction of the final energy demand due to economic crisis and lost export partners caused reductions in emissions of harmful pollutants to the atmosphere. Large share of nuclear in the primary energy budget also softens the problem.

But there still exist point sources of the atmospheric pollution - power plants and boiler houses burning high sulphur fuel oil, as there is no cleaning equipment installed at these plants. The energy sector is also responsible for a considerable amount of NO_x emissions.

Nuclear power plant at Ignalina may cause a potential radioactive contamination around it and also threatens with a possible serious accident. Problem of the nuclear waste temporary storage is solved. It is stored on site in concrete casks for the fifty year period.

Lakes used as coolers for the power plants are affected by thermal pollution and pollution from liquid effluents.

Every year about 200 thousand tonnes of hazardous waste is accumulated in the country. This waste according to the requirement of the Environment Protection Ministry (EPM) should be kept within the boundaries of the plant. In 1993 the Programme for treatment of the hazardous waste was be approved by the Government. All hazardous waste according to this Programme should be classified and monitored.

The National Environment Protection Strategy was prepared and approved in 1995. It defined the main priority goals in relation to the environment protection:

- reduce NO_x emissions from stationary sources;
- reduce SO_2 emissions from stationary sources;
- reduce CO_2 emissions;
- reduce dust emissions, etc.

In order to achieve these goals it is necessary:
- to start a process of legal reform in environment protection;
- to revise existing standards and ensure their compliance with the EU standards;
- to encourage the collection and use of environmental data and information as an environmental tool.

5.2 Instruments of regulation

5.2.1 Legal framework

The Environment Protection Law was passed in January, 1992. The Law acts as a framework statute to facilitate the pass of additional necessary legislation, and establishes the general principles for environmental protection. The main objectives of this Law are:

- to ensure ecologically sound environment;
- to reduce the negative effect of human activities upon the environments;
- to guarantee the right of Lithuania's inhabitants to a healthy, secure and clean environment as well as ensuring the preservation of biological diversity and ecosystems.

In 1996 the Government of Lithuania approved the National legal harmonisation programme, it envisages preparation of some environment protection regulating pieces of legislation in compliance with the EU directives. Until a unified legal framework is established, relations in the environment protection area are regulated by the constitution, environment protection law, Law of local governments, Law of the hazardous waste treatment a. o.

5.2.2 Administrative instruments

In Lithuania traditional administrative regulation measures are used: permissions, limits, norms, prohibitions, restrictions. One of the main measures is a normative value of the emissions of harmful pollutants. Temporary allowable emission value or maximal allowable emission value is fixed when issuing corresponding permissions. Two types of normative values are determined having in mind bad technical and technological state of the industrial units. Until 1991 only maximal allowable pollution values were fixed for the sewerage water output to the water reservoirs, these values were never exceeded. Only few polluters were granted with the permissions for the emission volumes to the atmosphere, and hazardous waste treatment was not regulated at all. Therefore in 1991 the former Environment Protection Department approved 'The order to issue permissions to use natural resources and determine normative value of pollutants emitted into the environment'. It obliged all legal and private persons executing any economic activity to obtain permissions for the natural resources use, and these permissions defined limits of the natural resources use as also normative values of the harmful emissions. Permission is issued with an agreement of the local municipality, it consists of the following parts:

- setting up the water consumption limits;
- emission of pollutants with the sewerage water;
- emission of pollutants into the atmosphere;

- treatment of hazardous waste;
- use of natural resource, etc.

There are limits of the natural resource use determined in the permissions, according to the type and volume of the production. On the same basis emissions of the pollutants are calculated, i. e. technically and economically grounded normative values. The approved order on the issuing of permissions for pollution describes the case when the user of the natural resources is not able to follow the maximal allowable normative pollution value. In this case a temporary allowed order for the pollution (of water, air) is set in the permission, based on the existing production technologies and emission treatment possibilities. In any case the maximal allowable pollution level is determined together with the necessary environment protection measures (new treatment technologies, sanitary protection zones, etc.). Environment protection measures are determined in the permissions, implementation of these measures allows to reduce anthropogenic impact on the environment. Every polluter must present a balance of materials used in the production process and of the corresponding residuals. It allows to evaluate volumes of the toxic materials. The permission determines conditions for storage, disposal and utilisation of the waste.

In 1993 the Government of Lithuania approved normative values for emissions from steam and water heating boilers. It was also planned new stricter standards to be valid since January, 1996 (Table 1). Polluters exceeding allowed limits of the natural resource use or of emissions may be sued or fined. If polluter does not follow other conditions set in the permission (does not monitor volumes of the use resource, does not analyse quality of emissions, does not implement environment protection measures, allowing to reach the maximal allowable emission limits), may be fined or even limited ones activity. The system of permission is one of the main instruments in reduction of the negative anthropogenic impact on the environment, in planing the priority measures to reduce pollution and forecast development of the environment. One of the main legal acts describing the administrative instruments is the Administrative legal volume code.

5.2.3 Economic instruments

Experience from various countries has shown that the best result in environmental regulation is achieved when combining administrative and economic instruments. This combination is especially efficient for economics in transition - economic instruments are not able to follow inflation and more efficient are administrative measures: permissions and prohibitions.

The main economic instruments are:

- natural resource and pollution taxes (this is probably the main and most widely used instrument);
- subsidies and loans;
- custom taxes and regulated prices;
- economic stimulation, compensation and fines.

Economic instruments are directed towards economic interests of the natural resource users and makes them to correct their behaviour seeking the maximal economic benefit or minimise losses to be consistent with the benefit to the environment.

Law on Royalties (taxation of the natural resources use) was passed by the Lithuanian Parliament in March, 1991. The purpose of the law was to increase responsibility of the resource use seeking that these resources were used efficiently and economically. As all natural resources are national assets and exclusively belong to the state, the Government should earn an income corresponding to the used resource price. The Law took into account only one component of the natural resource tax - cost of the state on prospecting, analysis and evaluation of the natural resources deposits. But recovery of only these costs does not stimulate efficient use of the natural resources. Therefore the Law on taxes for the natural resources use was amended with an additional tax (tax for the natural resources). This tax reflects the real characteristics, natural conditions of the resources. Calculation of these taxes was improved and simplified in 1996. The existing taxation system was revised, and tax calculation was amended with a new component - value of the resource. New tariffs were set based on the information about the market value of certain resources. Based on the resource price, viability of the plant and possible impact of the increased tax on the price increase the 5% of the resource price tax was fixed.

In April, 1991 the Parliament passed the Law on Taxes for the Environmental Pollution. It encouraged polluters to implement various environment protection measures and reduce a negative impact on the environment, as indexation of these taxes was not following inflation, especially in 1992 and 1993 when inflation reached several hundred percent, they did not stimulate reduction of pollution and played only a fiscal role - collection of revenues into the national or municipal budgets. The biggest impact of inflation was on the pollution taxes, as they were indexed lately. Indexation was lagging behind of the inflation until 1995. Besides it many of large companies were separated into small ones, and this created difficulties in issuing permissions and controlling emission levels. Consequently taxes for the pollution of the environment do not play the main stimulative role. Therefore pollution taxes are under revision, the main purpose of the revision is to simplify rather complicated tax calculation and payment system and increase the tax tariff in order to restore the main function of the tax - to stimulate polluters to reduce their pollution by installing new technologies or treatment of emissions. One of the main goals of the restructuring of the pollution taxation system - to implement principle 'polluter pays', i.e. polluters should recover losses caused by their pollution. In the new draft pollution taxation law the tax tariffs are set in connection with the goals of the environment protection, especially for the materials, important for Lithuania.

The other economic instruments are subsidies and loans allocated by the Government. Some subsidies or loans for environmental investments are a necessary short term measure as people are not able to finance such investments themselves. Lithuanian Government every year allocates some subsidies for the environment protection, this is traditionally used for the water treatment investment. Other environmental investment are not financed from the state budget (e.g. in 1993 there was 35,9 MLt allocated for the water treatment investments and only 0,95 MLt for the other environmental investments).

Indirect subsidies - reduction of the profit tax - are also used in the environment protection. Profit tax is reduced to 15% for the companies in charge of reprocessing and treatment of the waste or secondary raw materials. This approach is recommended for other environmental activities.

Preferential loans with a lower interest rate are planned for the environmental investments. But this measure is not offen used. In the case of a preferential loan, difference between the interest rates is covered by the Ministry of Finance. Nevertheless, a preferential loan is better than a subsidy, it makes municipalities (or other borrowers to use a loan more efficiently and control its implementation).

In order to restrict export of some natural resources (e.g. peat), custom duties may be imposed. In 1995 the Ministry of Environment Protection prepared a new order of custom duties differentiation for peat according to its place of destination.

Penalties are foreseen in the Royalties tax and Taxation of the environment pollution laws, these penalties exceed the tax several times when the given normative values are exceeded. Methodology for calculation of losses to the environment when breaking environment protection law was prepared in 1991.

Income collected applying economic and administrative measures is accumulated in

- the state budget;
- environmental funds of municipalities;
- state environmental fund.

It was decided to transfer all taxes for the natural resources use and 30% of taxes for the pollution to the state budget, the rest - to the municipal budgets. Only for the biggest Lithuanian power plant this proportion is different - 90% to the state budget, and 10% - to the municipal budget.

There are 56 municipal environmental funds in Lithuania, and these funds cannot accumulate enough of money for investments. In order to accumulate money for environment protection investments in autumn, 1996 the new Lithuanian environment protection investment fund was established. It should help to solve environment protection problems in the private sector. Environmental investments from the state budget, environmental funds and other sources are coordinated with the foreign assistance (e.g. the EU PHARE programme). Some investments are financed on the basis of bilateral agreements with Western countries. The largest foreign investment for environmental projects come from Denmark, Sweden and Finland.

Environmental monitoring system is crucial for an efficient use of the economic instruments of environmental regulation. This system would allow to plan changes of the taxation system, and improve the total environmental control mechanism. International experience of the economic instruments use may be transferred into Lithuania. Economic instruments, as permission allowances trade, environment insurance, etc. may be analysed and used.

5.2.4 Environmental expertise

Environmental expertise is an evaluation of impact of various activities on the environment. In 1990 an order of this expertise for a new construction was approved. At the same time the more strict requirements for the design and siting were set. Design projects of all units causing some negative impact on environment should be approved after their expertise is done at the Ministry of Environment Protection. The environmental expertise consists of the following stages:

- environmental analysis of the site;
- feasibility and economic concepts;
- design;
- monitoring and control.

The Parliament of Lithuania passed the Law on the evaluation of environmental impact of the planned economic activities. Regulations following this Law will implement new evaluation principles, corresponding to the international requirements fixed in the EU directive.

5.2.5 Environmental standards

Environment protection in Lithuania is tightly connected with the growing number of international agreements, rules and conventions. International relations, market conditions set limits for the difference of environmental requirements. At present, it is very important to harmonise normative documents on environment protection with those acting in the EU, or world-wide. Environment quality standards are under revision taking into accounts recommendations of the EU and other international institutions. Serious environmental impact of the energy facilities sets a priority requirement to determine the emission standards. Therefore new standards for emissions from the energy facilities is under preparation. It will determine the sulphur dioxide and nitrogen oxide normative values based on the best environmental experience and best existing technology, taking into account loads of sulphur and nitrogen compounds on the Lithuanian water and continental systems. For the preventive regulation of the emissions from the energy sector oil products quality standards are also important. They are consistent with the EU requirements and limit sulphur and hydrocarbon emissions.

5.3 International co-operation in environmental regulation

Efficient environmental policy is impossible without implementation of an international experience, without international cooperation. EPM paid serious attention to the collection of an environmental literature, participation in activities of international environmental organisations, joining international treaties. Some international conventions are implemented, some others require creation of the legal framework, some preparation work. In autumn, 1993 Lithuania joined the Geneva Convention on transboundary transfers of the atmospheric pollutants. It should be stressed that volume of emissions of pollutants listed in the Convention is decreasing. SO_2 emissions were reduced by 58% in comparison with the 1980 level, requirement of the Convention is only 30%. In 1990 the Programme for reduction of the nitrogen oxides was prepared, it is based on implementation of realistic measures (new burners). These burners were implemented at the Lithuanian Power Plant, at heating plants in Utena, Panevezys, etc. It resulted in 30% reduction of the NO_x emissions. Investments of 55 thousand Litas at the Vilnius CHP in 1992 enabled to reduce NO_x emissions by 40%. In general, NO_x emissions were reduced by 42.5% in comparison with that in 1987 (Convention required to reduce NO_x emissions in 1994 to the 1987 level). Emissions of volatile organic compounds (VOC) were reduced by 45% (Convention required to reduce VOC emissions in 1994 by 30% in comparison with that of 1988).

In 1992 Lithuania signed the Rio de Janeiro Convention on the Climate Change. This Convention was ratified by the Parliament in 1995. Therefore Lithuania monitors emissions of the greenhouse gases. The main greenhouse gas is carbon dioxide, and the main source of its generation is the energy sector.

Effective way in solving international environmental problems goes through good contacts with environmental institutions from other countries, implementation of the foreign experience, open discussions of the main issues in international forums, monitoring and control of the state of surrounding environment. Environmental standards and normative values of emissions are under preparation together with the Western experts. The system for the treatment of hazardous waste is prepared together with the Danish experts. US Environment protection agency organised courses on the assessment of the environmental impacts. Treaty on the environmental cooperation between Sweden and Lithuania was signed in 1992 and is successfully implemented.

References

Statistical Yearbook of Lithuania (1994 – 1997) (1994) Department of Statistics, Vilnius.
The National Energy Strategy (1994), Ministry of Energy Vilnius.
Jankauskas, V. (1996), 'Forecasts of energy prices', in *Lithuanian, English summary - Energetika',* No. 2, pp 21-30.
Lithuanian energy sector in 1996 (1997), Lithuanian Energy Institute.
Lithuania's own energy resources study (1994), Lahmeyer International, COWI consult and Lithuanian Energy Institute.
INPP Safety Improvement Programme (April 1998), Status Report.

Index

abatement techniques 25
abatement technology 30
acid rain 3, 8, 11
Advanced Gas Cooled Reactor (AGR) 52
atmospheric pollution 151, 163
autogenerated electricity 79

Baltic Dispatch Centre 151, 161
Baltic Gas Study 162
Baltic States 163
bio diesel 147, 153, 154
biofuel 35, 37, 144, 150
biogas 35, 37, 42, 74, 76, 145, 146, 147, 154, 151, 152, 153
biomass 34, 35, 37, 39, 41, 42, 73, 77, 91, 129, 130, 131, 133, 134, 138, 143, 144, 145, 129, 132, 136, 137, 138, 142, 148, 149, 151
BNFL 53
British Energy 51, 53, 57, 65, 66, 67, 68, 69

carbon dioxide 9, 11, 83, 171
carbon tax 9, 12, 13
CCGT 83
central heating systems 117, 118, 124
CfDs 76
climate protection 120, 123
climatic situation 124

CO_2 33, 41, 54, 82, 99, 100, 120, 122, 164
coal 4, 6, 7, 10, 11, 12, 14, 19, 101, 104, 117, 118, 120
coal power 104
coal production 10
coal sector 12
concerns of natural protection 139, 143
custom tax 167

Danish manufacturers 136, 140
Department of the Environment 73, 84, 97
DGES 58, 75, 76, 77, 78
district heating 117, 118, 120, 122, 143, 152, 153, 149
domestic pumping stations 152
domestic waste 145, 151, 152
dust 25, 28, 143, 145, 164, 149, 151

ECAFW 87, 90, 91
Economic Community (EEC) 4
economic position 11
EdF 62
electrical net 104
electricity 6, 7, 8, 10, 12, 13, 14, 15, 19, 20
Electricity Act 1989 74, 76, 78, 98

173

electricity consumption 102, 103, 105, 118, 119, 121
Electricity Feed Law 136, 137, 138, 139, 140, 141, 144, 147, 140, 141, 142, 144, 145, 146, 149, 154
electricity grid 32
electricity market 52, 54, 56, 69
electricity price 12, 80, 132, 141, 142, 148, 135, 146, 147, 155
electricity production 104
electricity sector 148, 158, 159, 160, 161, 154, 155
emission 2, 3, 8, 9, 11, 13, 18, 22, 24, 25, 28, 31, 33, 41, 43, 55, 80, 83
emission limit 146, 153
emission problem 3, 144, 150
emission reduction 8, 141, 147
emission standards 30, 31, 145, 151
Energy Act 1983 74
Energy Committee 50
energy consumption 99, 101, 102, 105, 106, 107, 108, 109, 110, 112, 113, 114, 115, 116, 117, 118, 119, 120, 121, 123, 124, 125
energy crops 81, 87
energy efficiency 3, 6, 9, 13, 16, 18, 20, 22, 23, 24, 26, 28, 31, 46, 99, 100, 101, 103, 105, 106, 107, 109, 112, 113, 118, 119, 120, 121, 123, 156
energy equipment 155
energy industry 1, 15
energy law 137, 138, 139, 141, 146, 147, 154, 155, 157, 141, 142, 143, 144, 146, 152, 153

energy market 2, 5, 6, 7, 12, 15, 17, 20, 56
energy policy 1, 2, 3, 4, 5, 6, 7, 8, 9, 10, 12, 17, 18, 22, 80, 97, 150, 154, 155, 156, 157
energy potential 153, 154
energy production 3, 4, 10, 20, 134, 153, 154, 156, 137, 138
energy resource 2, 9, 20, 153, 154, 155, 156, 157, 172
energy sector 2, 4, 7, 8, 9, 10, 12, 20, 131, 140, 147, 150, 151, 155, 156, 157, 158, 164, 170, 171, 172, 133, 145, 154
energy security 34, 38, 52, 53, 54
energy sources 74
energy supply 5, 19, 99, 100, 108, 151
energy surplus 6
energy tax 148, 155
Energy Technology Support Unit 81
England 72, 74, 76, 79, 83, 88, 95, 96, 97
environment 1, 3, 7, 8, 9, 11, 14, 16, 22, 23, 28, 29, 44, 46, 47, 51, 52, 54, 57, 61, 70, 136, 143, 140, 149
environment insurance 169
environment protection 156, 164, 165, 166, 167, 168, 169, 170
environment protection law 156, 164
environment protection ministry 164, 170
environmental benefits 14, 16, 134, 138

174

environmental bonus 148, 155
environmental consequences 13
environmental costs 4
environmental damage 128, 129
environmental demands 104
environmental energy 134, 138
environmental expertise 169
environmental externalities 47
environmental impact analysis 140, 145
environmental policy 1, 8, 9, 170
environmental problems 54, 153, 171
environmental protection 3, 4, 6, 8, 9, 17, 20, 22, 25, 27, 29, 30, 32, 44, 47, 55, 138, 146, 164, 142, 152
environmental protection law 139, 143
environmental regulation 51, 69, 150, 151, 154, 166, 169, 170
environmental requirements 152, 163, 170
environmental standards 155, 163
environmental tax 81
EU Electricity Directive 159, 160
EU energy policy 1, 7
EU Gas Directive 163
Europe 105
European Coal and Steel Community (ECSC) 4, 11

FBC 91, 92
Federal Act on Immission Protection 123
federal environmental law 146, 153
Federal Office for Environment 119
final energy 102, 105, 106, 107, 108, 109, 112, 114, 115, 118, 119, 120
financial subsidies 147, 153
flat 116, 125
fossil energy resources 99
fossil fuel levy 50, 51, 56, 57, 59, 60, 61, 62, 64, 65, 66, 72, 74, 75, 78, 79, 80, 81, 82, 89, 94, 95, 96
fossil fuels 9, 11, 128, 148, 129, 155
fuel cells 81
fuel diversity 34, 47, 53
fuel wood 129, 130

garbage system 145, 151
gas 5, 6, 7, 12, 17, 130, 131, 132, 133, 134, 135, 138, 143, 145, 146, 147, 131, 132, 133, 134, 136, 138, 139, 142, 149, 151, 152, 153
gas market 163
gas prices 135, 139
gas sector 158, 163
gas storage 162
gasification of wood 143, 149
GDP 23, 24, 26, 27, 30, 46, 101, 102, 103, 105, 127
geothermal 34, 35, 36, 42, 73, 76, 81
geothermal energy 129, 134, 141, 129, 138, 146
geothermal water 154
Germany 99, 101, 102, 103, 104, 105, 106, 108, 109, 110, 111, 113, 115, 116, 117, 118, 119,

120, 122, 123, 124, 125, 128,
129, 130, 131, 132, 134, 135,
136, 138, 140, 142, 148, 128,
129, 130, 131, 133, 134, 135,
137, 138, 140, 142, 145, 148,
154
global warming 8, 11, 15
Green Paper on Renewables 16
greenhouse effect 8, 12, 99
greenhouse gas 171
group interests 45, 46

hard coal 32, 70
heating energy consumption
 117, 125
heating system 152
hot dry rocks (HDR) 73
household sector 115, 118, 119
hydro 153, 154, 157, 159
hydro power 35, 104, 129, 131,
 140, 141, 129, 130, 133, 145,
 146
hydrocarbon emission 170
hydroelectric power 81
hydrogen 134, 137
hydro-storage plants 74

Ignalina Nuclear Power Plant
 150, 151, 152, 155, 158, 161,
 164
IKARUS- project 131, 133
imports 5, 10, 19
industrial waste 83, 85, 88, 90,
 93, 94
inflation 166, 167
instrument 24, 29, 30, 31, 48,
 147, 154
internal consumption 103, 104

internal energy market (IEM) 7
internal market 9, 15, 17
inter-sectorial change 107
intra-sectorial change 107
investment subsidy 136, 137,
 140, 142

joint implementation 45

laissez faire 29
landscape protection 138, 139,
 140, 143, 145
Law on Pollution Taxation 157
leviable electricity 78
levy 50, 57, 59, 60, 62, 66, 67,
 68, 69, 70
Lithuania 150, 151, 152, 153,
 154, 155, 157, 159, 161, 163,
 165, 166, 168, 169, 170, 171,
 172
Lithuanian Gas 158, 162, 163
Lithuanian Power System 151

Maastricht Treaty 17
market liberalisation 3, 4, 9
market prices 80
mini-hydro 15
Ministry of Environment 27, 29,
 30, 31, 48
multiple regulation 1, 10

National Audit Office 73, 97
national autonomy 13
National Control Commission for
 Prices and Energy 155
National Energy Strategy 152,
 155, 160, 161, 172

National Environment Protection Strategy 164
natural gas 11, 19, 20, 152, 162, 163
natural protection 138, 139, 141, 142, 143, 146
natural resource 156, 157, 165, 166, 167, 168, 169
negative externalities 4
new business strategy 69
new federal states of Germany (NFS) 103, 104, 107, 108, 111, 116, 117, 118, 122
nitrogen oxide 170, 171
non-fossil fuel obligation 10, 15, 57, 58, 59, 68, 70, 72, 73, 74, 75, 76, 78, 79, 80, 83, 85, 86, 87, 88, 90, 91, 92, 93, 94, 95, 96, 97
non-fossil purchasing agency 58, 59, 64, 67, 75, 76, 77, 78, 83, 84
NO_X 25, 33, 41, 54, 164, 171
Nuclear Electric 50, 51, 52, 58, 60, 61, 64, 65, 66, 67, 68
nuclear energy 53, 54, 55, 56, 104, 105, 128, 135, 148, 129, 139, 155
Nuclear Energy Law 155, 156
nuclear fuel 14
nuclear industry 76, 96
nuclear obligation 75
nuclear power 3, 14, 15, 19, 50, 51, 52, 53, 54, 55, 56, 57, 61, 63, 64, 66, 68, 69, 76, 79, 82
nuclear power industry 76
nuclear power stations 51, 52, 56, 68, 70
nuclear station 53, 66

nuclear technology 14, 55

oil 3, 5, 6, 12, 17, 19, 129, 133, 135, 143, 130, 137, 139, 149
oil harbour 158
oil heating systems 118
oil industry 152
oil pipeline 158
oil price crises 120
old federal states of Germany (OFS) 104, 105, 108, 109, 116, 117, 122
OPEC 6, 99
over-subsidy 148, 155

passenger transport 110, 111, 112, 113, 114
peat 153, 168
PESs 75, 98
petroleum products 152
photovoltaic 129, 130, 132, 133, 134, 137, 141, 142, 129, 130, 131, 134, 135, 136, 138, 142, 146, 147
Poland 22, 24, 25, 26, 27, 28, 29, 30, 31, 33, 34, 35, 36, 37, 38, 42, 43, 44, 47, 48, 49
policy instrument 75
pollution 8, 23, 25, 26, 27, 28, 29, 30, 32, 41, 46, 47, 48
pollution charges 30
pollution of environment 157
pollution tax 163, 167, 168
power generation 10, 12, 13, 17
power plant market 104
power plants 103, 104, 150, 157, 158, 159, 160, 163, 164
Primary Contract 76

primary energy 12, 19, 20, 25, 101, 102, 103, 105, 109, 127, 128, 131, 163, 129, 133
principal gases 11
private households 105, 115, 118
public debates 3
pyrolysis 91

radioactive contamination 82, 164
RD&D programme 73
RECs 75, 76, 77, 78, 79, 83, 86, 89, 98
Regional Electricity Companies 57, 58, 59, 60, 67
renewable energy 8, 15, 19, 20, 22, 33, 34, 35, 37, 38, 39, 40, 41, 42, 43, 47, 48, 61, 66, 67, 69, 72, 73, 74, 75, 79, 80, 81, 89, 128, 129, 130, 131, 132, 133, 134, 135, 139, 148, 149, 153, 154, 156, 157, 160, 128, 129, 130, 131, 133, 134, 136, 137, 138, 139, 144, 155, 156
renewable technology 73, 89
resource law 141, 146
Russian United Power System 161

sewage gas 74, 83, 85, 87, 94
sewage plants 146, 152
sewage water 145, 151
sewerage water 165
single buyer system 13
single electricity market 12
SO_2 25, 33, 41, 54, 164, 171
solar 15
solar cell producers 138, 142

solar collectors 129, 142, 143, 129, 148
solar district heating 142, 148
solar energy 34, 36, 39
solar power 73, 77
solar thermal energy 130
solar thermal energy 129
solid waste 26, 27
special nuclear levy 161
SSEB 52
standards for emission 145, 151
standards for emissions 143, 147, 149, 154
state budget 168, 169
State Energy Inspection 155
state intervention 47
State Nuclear Power Safety Inspectorate 155
stationary sources 28
subsidy 11, 33, 37, 38, 44, 50, 55, 59, 136, 148, 167, 168, 140, 155
subsidy mentality 76
sulphur dioxide 13, 26, 28, 45, 170
sulphur fuel 11
sulphur heavy fuel oil 150, 152, 163
sustainable development 100

tax reductions 157
Taxes for the Environmental Pollution 167
thermal pollution 164
third Framework Programme 14
third party 160, 162, 163
third party access 13, 40
third party finance 16
third world countries 135, 139

Trade and Industry Select
  Committee 52
transformation sector 102
transition 22, 26, 28, 29, 30, 46,
  48, 49
transmission network company
  162
transport sector 109, 110, 112,
  114

UCPTE 161
UK 51, 52, 53, 54, 55, 56, 59,
  62, 63, 64, 66, 68, 69, 70, 73,
  74, 79, 80, 81, 83, 94, 96, 97

Wales 72, 74, 76, 79, 83, 88, 95,
  96, 97
water consumption 165
water law 140, 141, 144, 146,
  145, 146, 150, 152

water power 77
water protection 142, 148
wave power 73, 77
White Paper 17
wind 15, 141, 146, 149, 146,
  152, 156
wind converter 131, 138, 139,
  133, 143, 144
wind power 129, 137, 139, 140
wind electricity 137, 141
wind energy 136, 138, 139, 142,
  154, 140, 142, 143, 144, 147
wind power 35, 36, 40, 42, 73,
  74, 77, 81, 84, 85, 87, 88, 91,
  93, 95, 96, 104, 133, 136, 137,
  138, 130, 137, 140, 141, 142,
  143, 144, 145
wind technology 73, 87, 94
wood 153, 157
wood industries 153

Trade and Industry, Select
  Committee 32
transformation sector 102
transition 22, 26, 28, 29, 30, 46,
  48, 49
transmission network company
  162
transportation 109, 110, 112,
(114)

UCPTE 161
UK 71, 52, 53, 54, 55, 56, 59,
  62, 63, 64, 66, 68, 69, 70, 71,
  74, 79, 80, 81, 82, 94, 96, 97

Wales 72, 74, 76, 79, 83, 88, 95,
  96, 97
water consumption 16
water, low 139, 141, 143, 146,
  147, 148, 150, 152

water power 27
water protection 142, 148
wave power 75, 77
White Paper 17
wind 15, 141, 145, 146, 149,
  152, 156
wind conveyors 131, 138, 139,
  135, 161, 144
wind towers 129, 137, 139, (140)
wind electricity 137, 141
wind energy 136, 138, 139, 142,
  156, 140, 142, 143, 144, 147,
wind power 35, 36, 40, 42, 73,
  74, 75, 81, 84, 85, 87, 88, 91,
  93, 95, 96, 104, 132, 136, 137,
  138, 150, 157, 160, 141, 142,
  145, 144, 145
wind technology 75, 87, 94
wood 152, 157
wood industries 153